The Child's Understanding of Number

The Child's Understanding of Number

Rochel Gelman and C. R. Gallistel

Harvard University Press
Cambridge, Massachusetts, and London, England

To Adam, who counts

10 9 8 7 6 5 4 3

Library of Congress Cataloging in Publication Data
Gelman, Rochel.
The child's understanding of number.

Bibliography: p.
Includes index.
1. Number concept. I. Gallistel, C. R. II. Title.
[QA141.15.G44 1986] 372.7′2′044 85-27506
ISBN 0-674-11636-4 (cloth)
ISBN 0-674-11637-2 (paper)

Preface, 1986

This book was written with two audiences in mind: those interested in cognitive development in general and those interested in the psychological foundations of mathematical thought in particular. Hence, it can be used in courses in cognitive development, the psychology of mathematics, and early mathematics education.

Chapters 1, 2, and 3 highlight the problems of standard accounts of early cognitive development and introduce the view that preschoolers have greater cognitive capacities than previously thought. Especially suitable for a course on cognitive development, these chapters focus on the difficulty of interpreting data that characterize preschoolers as unable to pass the same tasks as older children. We argue that the immediate success of appropriately designed training studies implies the existence of some prior understanding of basic concepts in the domains being trained. Otherwise the child would not be able to assimilate the training. However, such indirect arguments are just that. Chapter 3 surveys studies that were designed with the young child in mind and that demonstrate directly an early understanding of some fundamental concepts in the domains of communication and classification. The next seven chapters may be read as a case study in early cognitive development, using number concepts as the case. The advantage of this program is that it lends itself to a precise analysis of the variables which contribute to competence.

Chapters 11 and 12 focus on the theoretical implications of the findings regarding what preschoolers can do in the numerical domain, with Chapter 11 most relevant to courses stressing formal models of cognitive development. The rest of the book fits well into courses on early cognitive development, the role of structural constraints on learning, and theories of development.

Appropriate for a psychology of mathematics course, Chapters

5–12 serve to organize questions about the order in which arithmetic concepts develop; the role of counting, as opposed to logic, in the young child's ability to represent the numerical value of sets and reason arithmetically; the way mathematics informally learned by children naturally differs from mathematics learned in formal settings; and what children have to learn in order to expand their knowledge of arithmetic. The book can be used in conjunction with any of several recent, edited volumes on mathematical learning and knowledge in school-aged children (see the references at the end of this preface), such as Carpenter, Moser, and Romberg (1982); Ginsburg (1983); Heibert (1986); the Resnick and Ford (1981) text; and other research monographs, including Ginsburg's (1977), as well as Piaget (1952).

Finally, it is our experience that teachers and students of preschool curriculum find many useful ideas in Chapters 1–3 and 5–10. The emphasis on what preschoolers are capable of doing and knowing fits well with their own views that the very young actively engage themselves in the many different learning environments they are exposed to.

The work recounted in this book is brought up to date in the following papers: Gelman (1982a) develops the argument that the kinds of numerical abilities described in this book are universal. This paper, like the book, concentrates on preschool children's conceptual competence, those principles which constitute the core of their understanding of the domain of numbers. Greeno, Riley, and Gelman (1984) and Gelman and Meck (1986) turn to the question of the relation between this domain-specific conceptual competence and two other nonspecific kinds of competence—that at building up complex procedures from subprocedures (procedural competence) and that at assessing the general and idiosyncratic requirements of tasks (utilization competence). In both studies the authors argue that successful performance depends on the integration of these three distinct types of competence. In addition, Gelman and Meck illustrate how one can account for the relation between domain-specific constraints on cognitive development and the acquisition of general skills that span domains, thus providing a way to integrate two different theoretical approaches to cognitive development.

Our book argues that the conceptual competence of the young child consists of two sets of principles that guide behavior in the numerical domain. One set of principles defines and interrelates the operations of addition and subtraction and the numerical order and equivalence relations; the other set defines counting for the purpose of representing the numerosity of a set. The principles that define counting are the how-to principles discussed in Chapters 8 and 9. The abstraction and order-irrelevance "principles" have a different status. Whereas the how-to principles specify the properties of a proper count sequence and define what it means to count, the order-irrelevance and abstraction principles do not function this way; they are not an *active* part of the child's conception of number. Rather, they emphasize the absence of two widely presupposed constraints on the child's definition of counting. Young children do not restrict their counts to perceptually alike items. Nor do they use only those counting procedures for which they may have been reinforced. The absence of these constraints makes the how-to principles more general and powerful than they otherwise would be.

The only argument in the book we would want to change—as opposed to develop—in the light of subsequent work, is the one in Chapter 12, explaining why preschoolers fail the number conservation task. We proposed that, despite preschoolers' prowess in the numerical domain, they are unable to use a principle of one to one correspondence to establish and reason about equivalence. Gelman (1982b) tested this hypothesis and given her results, it is clear that there are some conditions under which children will recognize numerical equivalence on the basis of one to one correspondence (and hence conserve). Still, they clearly prefer to establish numerical equivalence through counting.

Recent References

Carpenter, T. P., J. M. Moser, and T. A. Romberg, eds. 1982. *Addition and subtraction: a cognitive perspective.* Hillsdale, New Jersey: Erlbaum.

Gelman, R. 1982a. Basic number abilities. In *Advances in the psychology of intelligence,* vol. 1, ed. R. J. Sternberg, pp. 182–205. Hillsdale, New Jersey: Erlbaum.

Gelman, R. 1982b. Accessing one to one correspondence: still another paper about conservation. *British Journal of Psychology* 73:209–220.

Gelman, R., and E. Meck. 1986. The notion of principle: the case of counting. In *Conceptual and procedural knowledge: the case of mathematics,* ed. J. Hiebert. Hillsdale, New Jersey: Erlbaum.

Ginsburg, H. P. 1977. *Children's arithmetic: the learning process.* New York: Van Nostrand.

Ginsburg, H. P., ed. 1983. *The development of mathematical thinking.* New York: Academic Press.

Greeno, J. G., M. S. Riley, and R. Gelman. 1984. Conceptual competence and children's counting. *Cognitive Psychology* 16:4–134.

Hiebert, J., ed. 1986. *Conceptual and procedural knowledge: the case of mathematics.* Hillsdale, New Jersey: Erlbaum.

Resnick, L. B., and W. F. Ford. 1981. *The psychology of mathematics for instruction.* Hillsdale, New Jersey: Erlbaum.

Preface to the First Edition

This book deals primarily with the preschool child's conception of number and how that conception develops. This central concern is, however, embedded in a larger framework, namely, an overriding interest in early cognitive capacities and their relationship to subsequent capacities. The initial chapters develop this broader framework.

The account of the young child's number concepts presented in the central chapters grew out of research by Gelman and her students on the cognitive abilities of preschool children. The first few chapters constitute a general preamble to the specific topic, the nature and development of number concepts. The preamble motivates our presentation of the experimental work on number, which focuses on the task of uncovering what young children do have in the way of numerical concepts. After reviewing previous work on the young child's use and understanding of number and highlighting the need to investigate counting abilities in children of preschool age, we present a model of counting abilities. The model is used to account for the data we have collected on the way young children count. We then turn to a consideration of how young children reason about number, treating counting as an algorithm that creates the representations of numerosity employed in reasoning. Having covered the way preschoolers represent and reason about number, we compare and contrast formal descriptions of arithmetic with our description of what preschoolers do and do not do in dealing with number. This comparison sets the stage for our discussion of the what and how of development.

The theory of the preschooler's understanding of number that emerges from our work is designed to account for the considerable range of things that the child can do as well as for the equally considerable range of things the young child cannot do. We aim to convince the reader that number concepts are a rewarding area for developmental theorizing precisely because of the richly articulated experimental picture of the child's achievements and shortcomings in this domain. This point brings us back to the more general argument that a deeper understanding of cognitive development depends on more detailed and elaborate experimental delineation of the cognitive structures possessed by very young children.

We wrote the nucleus of this book in 1973–74 while on sabbatical leave at the School of Social Sciences, University of California, Irvine. Duncan Luce and Rochel ran a seminar on basic concepts in mathematics and their development. Rochel insisted that Randy come to the seminar. Randy had been kibbitzing for years on her work on number concepts: He has an interest in the historical evolution of mathematical ideas, particularly the modern formalist approach. The mathematical sophistication of most of the participants in the seminar encouraged us to try to state the ideas underlying Rochel's research in a more formal way, a way that allowed sharper comparisons with the formal structure of arithmetic. We were sufficiently excited by this venture to spend every morning for several months writing everything out. The process of joint composition was too intricate to describe but proved surprisingly enjoyable and free of friction.

When the sabbatical leave ended, Randy resumed his research and teaching in physiological psychology. Rochel continued to develop and write out her ideas about early cognitive development, thereby creating a beginning for the manuscript. She also conducted and wrote up her research, a portion of which became Chapters 8 and 9. Randy came to her aid whenever writer's block drove her to the verge of despair, helped with the editing of the first ten chapters as they began to take on final form, wrote Chapter 11, and collaborated in the writing of parts of Chapter 12.

We are grateful to the Guggenheim Foundation for the fellowship that supported Rochel during her sabbatical leave at Irvine and to the Center for Advanced Study in the Behavioral Sciences, where she was a Fellow in 1977–78, when we finished this book.

Chapter 10 is based on an earlier paper (Gelman, 1977). The research reported in this book was supported by NIHCD Grant No. HD-04598 and NSF Grant No. BNS-7700327 to Rochel; so were many of the preparation stages of the book. Funds for Rochel's fellowship at the Center came from NIHCD, NIMH, and the Spencer Foundation. Very special thanks are due the staff, parents, and children of the YWCA Chestnut House, YWHA Broad Street Nursery School, and the Penn Children's Center—all of Philadelphia—without whose cooperation this research would not have been possible. We are grateful to Merry Bullock and Sue Merkin, who ran the studies. We are indebted to Renée Baillargeon, John Baron, Merry Bullock, Kay Estes, John Flavell, Herb Ginsburg, Duncan Luce, Ellen Markman, Paul Rozin, and Jerrold Zacharias for their careful, critical readings of some or all of earlier drafts of the manuscript. It has been a pleasure to work with our editors Eric Wanner and Camille Smith at Harvard University Press. Finally, special thanks go to Isabelle Friedman, who did the lion's share of the typing and retyping over the last four years and who made our life easier in innumerable ways, which can be imagined only by those who have been privileged to work with a first-rate secretary.

Rochel Gelman
C. R. Gallistel

Contents

Figures

Tables

Chapter One

Focus on the Preschooler

Children of preschool age sometimes give surprising answers to apparently simple questions. Consider the Piagetian number-conservation task. An experimenter shows the child two rows of objects, say a row of 10 flowers and a row of 10 vases. The two rows are arranged one above the other so that the one-to-one correspondence between them is obvious. For example, each flower may be directly above a vase. The child is asked whether there are the same number of flowers as vases. He usually answers yes. So far, not surprising. Next the experimenter, with the child looking on, spreads out the row of flowers so that it is longer than the row of vases. Again the experimenter asks whether there are the same number of flowers as vases. The preschool child almost invariably answers no, there are more flowers than vases.

The young child seems to believe that one can increase the number of items in an array simply by spreading the array over a larger area. So unexpected is this result that many parents, upon hearing of it, exclaim, "My 4-year-old would not do that!" They then leave for home with an experimental glint in their eye. When subsequently encountered these parents almost invariably report, with a mixture of puzzlement and interest, that their 4-year-old also behaves as predicted. Indeed, the failure of children younger than 5 to conserve, that is, to say "Yes, there are still the same number of flowers as vases," is one of the most reliable experimental findings in the entire literature on cognitive development. It has been reported hundreds of times, in a wide variety of cultures, often by experimenters who were initially incredulous. A few years later, at the age of 7 or 8, the child's answers match those of adults. By then the child is contemptuous of this younger siblings, who give such "ridiculous" answers to simple questions.

Nor is number conservation the only simple task that makes the preschooler appear cognitively inept. He seems, for example, to be unable to handle hierarchical classification. He cannot keep straight which is the superordinate class and which is the subordinate, particularly when it comes to quantitative questions. If one shows the preschooler a picture of some flowers, say six roses and two daisies, and asks, "Are there more flowers or more roses?" the child responds, "More roses" (Inhelder and Piaget, 1964).

When asked to describe an object in a way that will enable another person to identify which of several objects it is, the preschooler gives egocentric descriptions, that is, descriptions that represent idiosyncratic reactions to the object and are therefore of little value to a listener (Glucksberg, Krauss, and Weisberg, 1966). Consider the preschooler who describes a randomly shaped pattern as "Mommy's hat."

The list of further ineptitudes is long. The preschooler cannot treat two colors as belonging to the same dimension (Kendler and Kendler, 1962); cannot generate mediators when memorizing paired associates (Reese, 1962); cannot deal with part-whole relations (Elkind, Koegler, and Go, 1964); does not rehearse what he is later going to have to recall (Flavell, 1970); does not readily ignore irrelevant information (Gibson, 1969); has little metalinguistic awareness (Gleitman and Rozin, 1973a); and so on. (See White, 1965, for further examples of preschoolers' failures to pass simple cognitive tests.)

The cognitive shortcomings of the preschooler are many and well documented. To some extent the study of cognitive development is interesting precisely because the preschooler's performance is so remarkably different from that of the somewhat older child and the adult. Any account of cognitive development must deal with these differences.

Still, we see a danger in dwelling too much on the ineptitude of the preschooler. It is an easy step from these observations to the theoretical accounts that assume the complete absence of a variety of underlying cognitive capacities in the preschooler. Depending on our theoretical bias, we might say that the preschooler is perception-bound and therefore unable to think logically (Piaget, 1952) or symbolically (Bruner et al., 1966); that he fails to use the second signal system (Luria, 1961); that he is unable to form s_g-r_g mediators (Kendler and Kendler, 1962); that he has a "primitive" mind (Werner, 1957); that he thinks associatively but not cognitively (White, 1965; Jensen, 1969); that he lacks concrete operations and is egocentric (Piaget, 1926); and so forth.

Whatever the theoretical framework, the overriding tendency is to treat the preschooler's cognitive capacities, or lack thereof, in the light of those possessed by the older child. Knowing what the older child can do, we formulate a theoretical account of his capacities and slip into the position that such capacities are absent in the preschooler. Note what is going on here. Children of different ages are given the same task—a task that is assumed to be particularly well suited for testing a given capacity. The child who passes the test is said to "have" that capacity and the child who fails the test is said to "lack" the capacity. Since it is the older child who passes the test, the older child serves as the standard for the capacity in question. The definition of how the younger child differs from the older is given in terms of what capacity *the younger child lacks.*

This tendency to characterize the cognitive status of preschoolers in terms of what they cannot do is unfortunate for many reasons. Our first misgiving is methodological. In many cases, the child is said to lack cognitive principles of broad significance simply because he fails a particular test involving these principles. The belief that preschoolers lack number-invariance rules, for example, rests on their inability to perform a single task, the Piagetian number-conservation task. This task is certainly one test for the presence of number-invariance rules; it is difficult to see how a child could pass it without such rules (Gelman, 1972a). The converse, however, does not hold. Failure to pass the Piagetian number-conservation test cannot by itself be taken as proof that preschoolers lack number-invariance rules. Failure on a single test should not be accepted as proof of the null hypothesis under any circumstances. By definition any nontrivial cognitive structure will play a role in a variety of contexts. Before concluding that a particular structure is absent, we should at the very least test for it with a variety of different tasks, in each of which the structure plays a role.

Our second concern is with the theorizing that flows from this view of the preschooler as one who lacks a variety of cognitive capacities that 7- or 8-year-olds possess. Such a view of development makes life very difficult for the theorist who is interested in describing the process of cognitive growth from the preschool years through to middle childhood. The difficulty is that we are given all-or-none descriptive statements about cognitive development across a wide number of years. And all-or-none statements provide no constraints on the paths that might move an individual from the "none" stage at, say, 3 years to the "all" stage at, say, 7 years. As long as we are told that at one stage in cognitive development there is nothing and at the next stage there is everything, there are few constraints on our efforts to make guesses about the paths that might move a child from one stage to the next. This is an unacceptable state of affairs. Without some restrictions on the set of possible hypotheses, the search for acceptable hypotheses can continue indefinitely. And without constraints, how are we to choose among even a set of good guesses?

Choosing an appropriate theoretical framework would be easier if we had data on what preschoolers *can* do in the cognitive domain of interest. It is more difficult and interesting to explain progress from *x* to *y* when *x* is characterized no longer as "not *y*" but rather as having attributes of its own. This consideration leads us to prefer a developmental analysis that both compares and contrasts. When the preschooler is characterized only by his failures, the analysis rests on contrasts alone.

We do not mean to imply that cognitive developmentalists make random guesses about how children develop. The typical procedure involves analyzing the end product and then hypothesizing how it came about. Generally the theorist posits a particular skill, or rule, or strategy, or cognitive capacity that enables the child to conserve. Then the theorist argues that the child who fails to conserve lacks that particular skill, rule, strategy, or capacity. Armed with these hypotheses, the theorist turns experimentalist and conducts a training study. If the training is successful (say, when a child fails to conserve on a set of pretests but after some training manifests a general and lasting knowledge of conservation), the hypothesis is accepted. Gelman (1969) followed this tack in early work. She suggested that the child who conserves knows that he should attend to relevant and ignore irrelevant quantity dimensions. She then argued that if children who could not conserve were taught to shift their attention away from irrelevant dimensions to relevant ones, they would pass a conservation posttest. Her success in inducing generalizable and lasting knowledge of conservation led her, at that time, to accept her original hypothesis. Time has passed, however, and other successful training studies have been carried out, each one taken to support the hypothesis that generated it (see Beilin, 1971, for an excellent review). We agree with Beilin that the question is no longer "Can we train conservation?" but rather "Can we come up with a uniform consistent hypothesis about the development of conservation that can account for the various successes in training?" We suspect that the answer will be no as long as we confine our search to the analysis of success on the conservation task and the results of the training studies that have worked. What we need in addition is evidence of what knowledge and skills untrained preschoolers have.

This need for data on what untrained preschoolers can do shows up clearly if we look at the procedures used in successful training studies and the diverse hypotheses such studies supposedly verify. Some researchers interpret failure on the conservation task to mean that the child does not understand the quantitative terms *same amount, as much, more, less* (Braine, 1962). Some take the failure to mean that the child does not attend to the correct dimension of quantity. For example, Gelman (1969) assumed that the child failed to attend to the number of items in a row and focused instead on the length of a row. Some attribute the failure to a lack of understanding of reversibility, that is, a lack of appreciation of the fact that a particular transformation, such as lengthening, can be canceled by a complementary transformation, in this case shortening (Piaget, 1952; Wallach and Sprott, 1964). Some take the failure to mean a lack of measurement operations (Bearison, 1969). Others take it to reflect a lack of general quantitative experience (Kohlberg, 1968), and so on.

On the surface these hypotheses seem to be different, even mutually exclusive. We might expect each to lend itself to a distinctly different training paradigm. But do they? Gelman's training study (1969) provides an illustration of the ambiguous relationships between the various theories of why preschoolers fail Piagetian tests for number- or length-invariance rules and the training studies motivated by these theories.

Gelman worked with preschoolers who failed, on a pretest, four Piagetian tests for quantity-invariance rules. They failed the number-conservation test already described. They also failed a test for the invariance of lengths under displacement transformations. When two equally long sticks were presented one above the other with their ends lined up, the children said they were the same length. When one stick was placed so that its ends no longer lined up with the ends of the other stick, they said one of the sticks was longer than the other. They also failed a test for the invariance of liquid amounts. When two equal amounts of water were presented in beakers of identical dimensions, they said the amounts of water were equal. When one of the two containers was emptied into a narrower and taller glass, they said that the taller glass held more water. Finally, they failed a test for the invariance of a given amount of clay. When two clay balls of identical diameter were presented, they said the balls contained equal amounts of clay. When one of the balls was then flattened out, they said it contained more clay.

Gelman reasoned that these children were not focusing on quantity as such in these tests and set about training them to pay attention to the quantities in the number and length comparisons. The training involved the use of 32 problem sets with 6 trials in each set. Half of the problems trained for number concepts, half for length concepts. On each trial the children were asked which two of three rows of chips (or sticks) had the same number (or length), or which one had a different number (or length). On the first trial of each set the arrays were arranged so as to elicit a correct answer even if the child was attending to an irrelevant aspect such as length or density of the rows (in the number tasks) or which end projected beyond the other (in the length task). For illustration of these first-trial displays, see Figure 1.1. On trials 2 through 5 of the problem set, the children first watched while the experimenter rearranged the rows (or sticks) so that attending to the number-irrelevant (or length-irrelevant) aspect of the array would lead to an incorrect answer when the experimenter repeated the question (see Figure 1.2). Children in the experimental group were told whether their answers were correct or incorrect. (A control group did not get this feedback.) On the sixth trial, the children again watched while the experimenter rearranged the rows (or sticks) so that the irrelevant aspect was unlikely to exert any control over the response to the question (see trial 6 in Figure 1.2).

Since within each problem set Gelman created a conflict between "perceptual" and quantitative cues, she provided attention training. In retrospect, however, it is clear that she did more than this. Trial after trial the child was asked questions about *same* or *different* numbers and lengths; thus the child had an opportunity to learn the appropriate use of these quantitative terms. Since the child watched the experimenter produce a wide variety of transformations, the training procedure could likewise have taught the child about the relationship between transformations and given quantities. Since the quantity domain (length versus number) and the amounts within a quantity domain varied, the training procedure could also have provided general quantitative experience. Thus the procedure could have been motivated by any one or all of a wide variety of competing hypotheses about the child's source of difficulty on the original conservation task. Not surprisingly, other investigators have offered diverse interpretations of what Gelman's training procedure was about (see, for example, Brainerd and Allen, 1971).

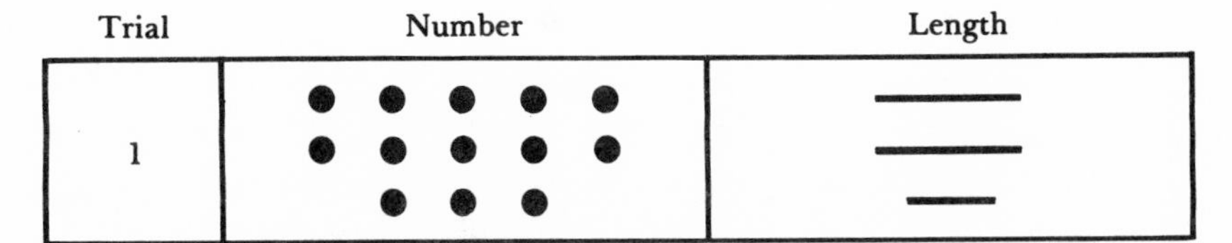

Figure 1.1. Schematic representation of the first trial in Gelman's training procedure. Examples are given for both number and length problems. In the number example, the top two rows have the same number. They are likely to be judged "the same," however, even if the child does not attend to number. Similarly, the top two lines in the length problem are likely to be judged "the same" even if the child does not attend to length. (From Gelman, 1969.)

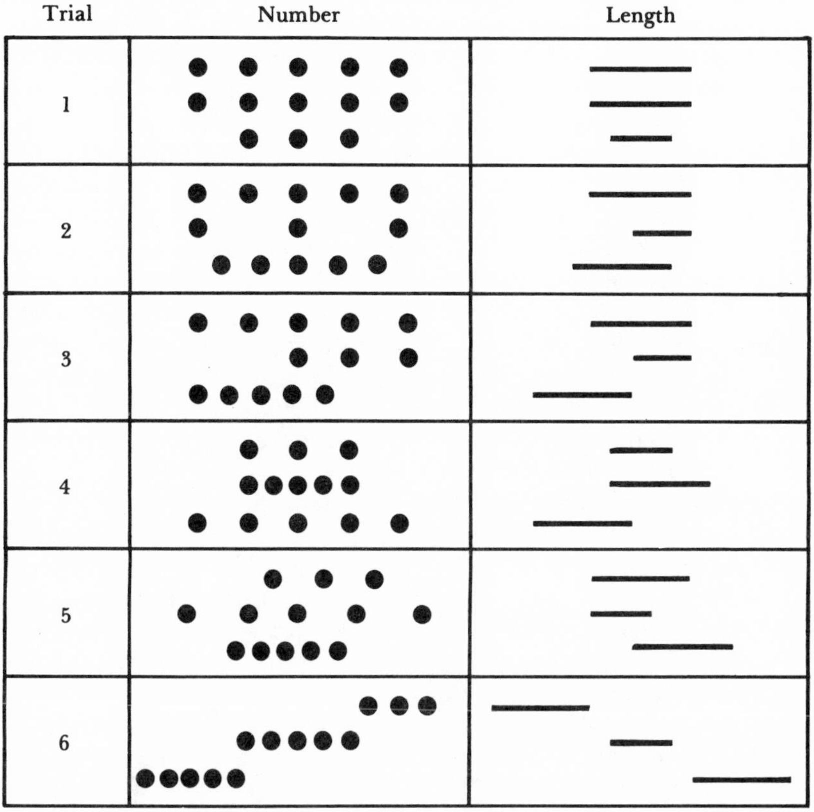

Trial	Number	Length
1		
2		
3		
4		
5		
6		

Figure 1.2. Schematic representation of a number problem set and a length problem set in Gelman's training procedure. The configuration in trial 2 of the number problem set would induce a child who was attending to the length of the rows rather than the number of dots to say (incorrectly) that the first two rows had the same number. Trials 3–5 are similarly calculated to mislead. On trial 6, however, the misleading irrelevant aspects of the display are minimized. Similar comments apply to the displays in the length problem set. (From Gelman, 1969.)

An inspection of other successful training paradigms yields a similar ambiguity. The Wallach and Sprott number-conservation training study (1964) was designed to teach nonconservers the principle of reversibility. This study involved a series of problems, each using two displays, one of *N* dolls and one of *N* beds. The numbers of items per array varied from one problem to another. Within each problem children were first shown that each doll fit in a bed. Then the dolls were removed from the beds and either the row of beds or the row of dolls was spaced farther apart or closer together. Children who said there were no longer as many dolls as beds were shown that each doll still had a bed, whether or not one row was more spread out than the other. The idea was to teach the children reversibility, that is, to teach them to predict that the dolls would fit back into the beds. But, as Wallach notes in a later paper (Wallach, Wall, and Anderson, 1967), this procedure could just as well have drawn attention to the misleading cues produced by displacement transformations. It also could have clarified the meaning of terms like *same* and *one too many* when used in reference to numbers.

Similarly, Bearison's training procedure (1969) could have taught children about measurement, or about the misleading effects of perceptual cues, variation in quantity, or the ways quantity terms are used. Since Bearison had children pour liquids back and forth between beakers of different sizes, he also gave the children an opportunity to observe that one transformation reversed the other.

What did the children learn in these various successful procedures? It is by no means obvious that they learned what the experimenters assumed they needed to learn in order to conserve (see Halford, 1970). We contend that it will be exceedingly difficult to decide what they actually learned without an independent assessment of what they brought to the task. What emerges after a child has had some training is the joint result of the training and whatever interpretative structures the child brought to the training. A mind with no interpretative structure cannot in any meaningful sense be trained, or so we claim. Following in the tradition of the British empiricists, many psychologists used to maintain that the only interpretative structure a human had to possess in order to be trainable by experience was the faculty of associating an input with another input or with an output. Today, however, most psychologists would concede that faculties or principles beyond that of simple association must be posited in order to understand what emerges from the encounter between the organism and its environment. In Piagetian terms, there must be a structure that assimilates the experience provided by training.

We are ignorant of the cognitive structures that enable preschoolers to assimilate their experience. We do not know, for example, whether preschoolers already understand reversibility and the effects of transformations. If we learned that untrained preschoolers do understand these principles, we could argue that the procedures (*a*) did not teach these principles, (*b*) taught something else, (*c*) brought out the role of reversibility, (*d*) integrated the knowledge the child had with that he did not have, and so forth. In other words, we could begin to delimit the domain of hypotheses. Obviously, the more evidence we have about the preschooler's quantitative knowledge, the easier our job of explanation will be. Information about what tasks the preschooler can perform would make it easier to determine what makes him unable to perform others.

We believe that failure to experimentally uncover what the preschooler can do belies the commitment to a developmental approach and stands in opposition to our metatheories of development. Those who call themselves developmentalists often have to defend their belief in developmental research. This challenge frequently arises in the form of the question "Why bother working with kids when it's much easier to work on that problem with adults (especially college sophomores)?" We venture to guess the nature of the usual answer: "If you look at the end product and make up tales about its origins you are likely to be wrong. Furthermore, you might alter your theory of the end product if you knew the stages it passed through." To illustrate the first point, the frog: No amount of inspection of adult frogs will ever lead one to determine that they were once tadpoles. Evidence for the second point—that a knowledge of development will influence the theory of adult performance—is somewhat harder to come by, but it exists. Views of how to teach reading differ depending on whether they are generated from the skills of the good adult reader or of the child who is just beginning to read (Rozin and Gleitman, 1977). As we will show elsewhere in this book (Chapter 12), the view of the subitizing process as a simple perceptual means of apprehending number is altered when considered in the context of how 2- and 3-year-old children estimate the numerosity of small arrays. We can shift back to biology for an illustration of this second point. Our view of what parts of the adult brain are most intimately interrelated has been strongly influenced by studies of both the embryological and the phylogenetic development of brain structure. Without developmental studies, how would we know that spinal tissue derives from skin tissue? It seems inconsistent to argue for the necessity of research comparing children and adults while ignoring the need for comparing preschoolers and school-aged children. Yet until recently there has been little research directed at describing the cognitive capacities of preschoolers. (See Gelman, 1978, for a review of recent research.)

Uneasiness about having to guess what preschoolers can do by considering the capacities of older children is not unique to us. Flavell (1977) points out that we know a great deal about the cognitive inadequacies of the preschooler and, like us, suggests that it would help our theoretical efforts to have descriptions of the preschooler's capacities as well. He offers what we call a metatheoretical justification: He suggests that cognitive development in a particular domain can be characterized as a process of knowing more and more about that domain. This view that knowledge builds on knowledge is shared by Schaeffer, Eggleston, and Scott (1974), and by Klahr and Wallace (1976). This is but one possible metatheoretical view of cognitive development. There are others. It is striking how many of them are consistent with or even dictate the research strategy we advocate.

By *metatheory* we mean general theoretical assumptions about the way cognitive development proceeds—assumptions that hold across content areas and therefore are not tied to particular cognitive domains. Much has been written about the principles of cognitive development (see, for example, Baron, 1973; Flavell and Wohlwill, 1969; Flavell, 1971; Kohlberg, 1969; Turiel, 1969; Piaget, 1971). We focus here on the theoretical positions that point to the need to obtain empirical evidence about the skills of preschoolers.

One major type of metatheory is stage theory, which is often referred to as the cognitive-developmental view (see, for example, Kohlberg, 1969; Turiel, 1969). According to this view a child passes through qualitatively different stages of cognitive capacity with each stage assumed to depend in some way on the previous one. The manner in which later stages depend on earlier ones can vary. Turiel favors a linking hypothesis in which later stages incorporate earlier stages. Theoretically, a wide variety of relationships can hold between two stages. The earlier stage can be integrated hierarchically into the subsequent stage. A simple reflex can be incorporated into a more complex sequence through the development of what Gallistel (forthcoming) has termed selective potentiation and depotentiation. The earlier stage can serve as scaffolding—necessary for the construction of the later stage but not integral to its operation. The pupal stage of insect development constitutes such a scaffolding. The earlier stage can be transformed into the later stage, as in the development from tadpole to frog. And so on. Some theorists challenge the idea that cognitive development is best characterized in terms of stages (for example, Baron, 1973). If one accepts stage theory, however, the problem becomes one of choosing among the various possible relations between one stage and another.

The decision as to which of the potential relations holds between stage x and stage $(x + 1)$ in the development of a particular cognitive capacity rests ultimately with the data. That is, our observations about the accomplishments of stage x and stage $(x + 1)$ constrain our choice of a linking hypothesis. Knowing that a tadpole becomes a frog, we treat this developmental sequence as a transformational one. The tadpole is transformed into a frog, not incorporated into the frog as a subcomponent. By contrast, Piaget's suggestion that preschoolers possess independent quantitative functions leads to the idea that these functions are integrated to produce conservation-generating schemes (Flavell, 1977). Our knowledge of the capacities in stage x, in conjunction with our knowledge of the capacities in stage $(x + 1)$, helps us decide whether stage x serves as a catalyst, a component, a scaffold, or something else for stage $x + 1$. The view that cognitive development proceeds in stages cannot, on its own, show us how these stages are connected. Only careful description of the accomplishments of both earlier and later stages allows the theorist to choose from the range of possible relating functions.

Two points are implicit in this discussion. First, it is assumed that, in one way or another, the child's cognitive capacities at a given time depend on those that were there before. Second, the nature of this dependence is difficult to ascertain without descriptive data on the structure of early stages. Both points underscore, once again, the need for research on the workings of each stage. Nothing about the general theoretical enterprise suggests that it is sufficient to describe any stage by what it does *not* contain. We must describe in positive, not just in negative, terms the capabilities of each stage.

Parenthetically, the need for a description in terms of positive as well as negative attributes is explicit in Piaget's writings. It was recognized in his treatment of sensorimotor intelligence. But until recently (Piaget, 1975), such a description was not evident in his treatment of cognitive development as it proceeded from the completion of the construction of the belief in object permanence to the onset of concrete operations. His more recent work on the nature of the preschooler's understanding of quantity is a contrast to his earlier descriptions of the preschooler as prelogical, egocentric, and the like. Given Piaget's view that assimilation and accommodation serve as the mechanisms of cognitive development, it never sat well to think within the Piagetian framework of the preschooler in terms of a list of capacities he lacked.

Those who opt for a stage-theoretic account of cognitive development have a special obligation to discover what preschoolers can do. For stage theories contain an implicit or explicit assumption that preschoolers do things differently and not just that preschoolers can do fewer things. This is not to suggest that stage theorists alone have a reason to uncover early cognitive capacities. Learning theorists hold the view that development flows from experience. As experience accrues, responses are reinforced, stimuli are brought to control habits, and thus cognitive skills expand. The assumption is that preschoolers have no particular qualitative deficiencies in cognitive capacity, at least none that are unique to them. What distinguishes preschoolers from older children is their lack of experience. Such a theory suggests that it should be quite easy to find cognitive tasks that preschoolers can pass and that it should not be too difficult to develop learning situations that show these capacities being used in more and more advanced ways as the level of training advances. Yet, the "stageless" approaches to development have produced little along these lines.

Some maturation theorists likewise share the obligation to discover what preschoolers can do. According to one version of maturation theory, the underlying cognitive capacities of the preschooler, the school-aged child, and the adult share a common structure (see, for example, Fodor, 1972). The younger child's structure may be incomplete, not yet fully developed, but it is remarkably like the adult's structure. If this view is correct, there should be a subset of tasks that both young and old do well.

As developmental psychologists—from the standpoint of both methodology and theory—we are committed to the empirical investigation of the preschooler's capabilities. We should avoid the tendency to compile a list of what preschoolers cannot do that older children can do. This tendency amounts to working backward from the full-fledged showing of a capacity. We are all aware of the danger of proceeding this way: There is no guarantee that the end state embodies the earlier stages of development. The emphasis must be on a consideration of the earlier stages in their own right. We must look for skills young children have—at least as much as we look for skills they lack.

Chapter Two

Training Studies Reconsidered

We think the cognitive capacities of preschoolers have been underestimated. We do not deny that young children fail a wide range of tasks. Rather, we are wary about concluding that they lack the crucial capacities or skills assumed to be tapped by the tasks. A critical look at training studies illustrates the pitfalls of too readily drawing such conclusions.

Conservation

According to Piaget, children fail his conservation task because they are unable to treat quantity as invariant through quantity-irrelevant transformations. The child who cannot conserve is said to lack quantity concepts. A child who fails a number conservation task, for example, is characterized as one who has yet to develop a concept of number. We think that some of the studies designed to teach nonconservers to conserve raise serious doubts about such interpretations. Gelman's training study (1969) is one good example.

Gelman used a learning-set procedure that required children to respond to the relevant quantity dimensions and ignore irrelevant ones. As Figure 2.1 shows, when she gave nonconserving children learning-set training, they quickly reached plateau in the training phase. That is, when given feedback, they quickly came to respond to quantity as such when asked to choose the two of three arrays that contained the same or different quantities.

The rapid rate of acquisition provided food for thought. It seemed unlikely that children who lacked quantity concepts would acquire them from scratch in such short order. One feels somehow that fundamental concepts of far-reaching application are not learned from scratch in a few trials. For this reason, Gelman argued that the training served to uncover a preexisting ability or to bring into play a pre-

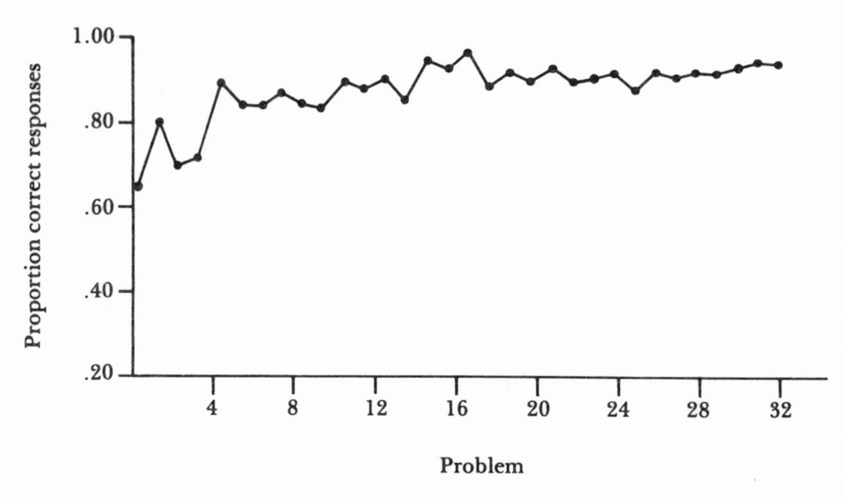

Figure 2.1. Learning-acquisition curve for subjects in learning-set training. (From Gelman, 1969.)

existing conceptual scheme rather than to develop from scratch the concepts of number and length and their invariance. Perhaps the training simply clarified what the experimenter wanted the child to do.

Two other results support Gelman's conclusion. Each problem set presented during training included one trial in which the three rows (or sticks) were arranged to exclude the use of irrelevant cues. If the children had no understanding of number or length, they should have responded at chance on this trial. The conceptually misleading aspects of the stimuli could not guide their behavior, and if they had no appropriate concepts to guide them they should have chosen at random. But they did not; indeed they usually answered correctly on these trials, indicating that they were able to make judgments on the basis of quantity as such.

Results of the generalization tests also strengthen the view that training uncovered preexisting concepts. After training on number and length, 55–71 percent of the children responded correctly on tests for the principles of liquid (water) and mass (clay) conservation. One way to account for this ability to generalize is to assume that the children had some preexistent understanding of quantitative invariance. Indeed such an assumption seems necessary. How else could children who had received no training on liquid and mass concepts perform so well on tests of these concepts?

It is important to make clear what we are and are not saying about the Gelman training study. We are not saying that it served no teaching function at all. Since the children initially failed to conserve, they must have learned something during training. We *are* saying that what they learned may very well not have been the fundamental principles that guide adult thinking about number and length. The rapidity with which training progressed and the range of problems to which it generalized suggest that Gelman was not engraving basic concepts upon *tabulae rasae.* More importantly, we conclude that successful training studies suggest that children already possess *some* conceptual scheme. Further experimental exploration is needed to determine what training does and does not add to this conceptual scheme.

A look at successful training studies in areas other than conservation will serve to buttress our argument. The character of each success suggests some preexisting skill, ability, or conceptual scheme. These training studies are a rich source of suggestions about what the untrained preschooler may already have rattling around in his head.

Transitivity

Bryant and Trabasso (1971) began what became a very lengthy and interesting study (Trabasso, 1975) of the young child's purported inability to make transitive inferences. They worked with children aged 4 to 6 years, an age at which children typically fail Piaget's test of transitive inference. In the Piagetian task the child is shown two of the possible pairings of three sticks of different lengths, for example, *AB* (with $A > B$) and *BC* (with $B > C$). He is then asked to determine which member of the third pair (*AC*) is longer, without being allowed to inspect that pair. Piaget interprets failure on such tasks as a reflection of the young child's inability to logically add the relations $A > B$ and $B > C$. Bryant and Trabasso suggested that this failure might derive more from the memory demands placed on the child than from the lack of a principle of transitive inference. To test their hypothesis, they devised a memory-training procedure.

Bryant and Trabasso showed children pairs from a set of five sticks (*A B C D E*) of different lengths and colors and taught them by means of a discrimination-learning technique which of a pair was longer (or shorter). The child was trained first on the *AB* pair, then on the *BC, CD,* and *DE* pairs. This phase completed, the child proceeded to the next phase, in which he encountered random pairs of sticks and thus had to learn a new series of discriminations. Two features of the training are noteworthy. First, the children never saw the actual lengths of the sticks; the bottoms of each pair of sticks were hidden in a box, and their tops protruded to the same height. Thus the children had to learn to associate the different colors with different relative lengths. Second, training required the children to respond to "which is longer" and "which is shorter" questions—a feature that served to highlight the comparative relations (Riley and Trabasso, 1974). After being trained, the children were tested, without feedback, on all 10 possible pairs of sticks. As in training, they had to rely on the color of the sticks when choosing one as longer or shorter. The test of their ability to make transitive inferences focused on the *BD* comparison and the critical adjacent pairs *BC* and *CD*. Children were not trained on the *BD* comparison. Further, during training the sticks of this pair as well as stick *C* were as often the longer as the shorter stick in an array. The percentage of correct responses on the *BD* comparison was considerably above chance, ranging from 78 percent to 92 percent. Success on the critical test pair *BD* was highly correlated with a child's ability to remember the values of the relevant adjacent pairs, *BC* and *CD*. The latter result is the major source of evidence supporting the Bryant-Trabasso hypothesis. Once again we are confronted with a training study that seems to redefine the nature of a young child's difficulties. By training the child to remember the critical information, Bryant and Trabasso uncovered a conceptual scheme, one that incorporates a rule of transitive inference. Thus, they discovered a domain of competence that had been presumed lacking in the young child.

Memory

Bryant and Trabasso's research suggests that the young child's difficulties with cognitive tasks do not all reflect a deficient cognitive structure. Some of the difficulty may derive from poor memory skills. Early work on the development of memory uncovered a variety of ways in which young children lag behind somewhat older children in memory tasks. When young children are asked to learn a series of items for later recall, they seldom if ever rehearse (Flavell, Beach and Chinsky, 1966). The failure to rehearse adversely affects their recall scores. A brief period of instruction and training in rehearsal suffices to get the nonrehearsers to rehearse when subsequently told to do so; and their recall scores improve (Keeney, Cannizzo, and Flavell, 1967). Yet these same children stop rehearsing once the experimenter stops telling them to. These findings led Flavell (1970) to conclude that the children were not deficient in their ability to rehearse and thereby aid their recall. Rather, he concluded, their problem was that they did not spontaneously think of rehearsing.

It is noteworthy that in the Keeney, Cannizzo, and Flavell study the children ceased to rehearse once the experimenter stopped asking them to do so. Similar findings have been reported often enough (see Brown, 1975b) for us to conclude that they are reliable. We think this finding is a bit odd; after all, when young children do rehearse, their recall scores improve. One possibility is that the children do not realize that rehearsal improves later recall. As far as we can tell, rehearsal-training studies do not provide feedback regarding the efficacy of rehearsing.

A study by Markman (1973) provides some indirect support for the idea that young children need to be *told* to monitor their output under different conditions. The Markman study was designed to determine whether preschoolers lack the ability to monitor their memory processes. Young children apparently think that they can remember rather large amounts of material. For example, they say yes when asked if they would be able to remember what was on as many as 10 pictures (Flavell, Friedrichs, and Hoyt, 1970). Markman found that a little experience or feedback about how many items they actually recalled improved the performance of kindergarten children on a memory-monitoring task. Many children who were judged to be poor memory monitors (that is, unable to predict how many items they would be able to remember) became accurate predictors after brief training. Furthermore, after training they were able to make differential predictions for items the experimenter judged to be easier or harder to recall. (Pictures of uncommon objects were judged to be more difficult to recall than pictures of common objects.) A follow-up test after two to three weeks revealed the "acquired" ability to be stable.

Markman commented on how readily she influenced the children's memory-monitoring skill. We suggest that the children needed to see the effect of memory monitoring on the accuracy of their predictions rather than to acquire the ability to monitor their memories. In any case, the literature on memory training calls into question the idea that the young child's poor performance on some standard tests of memory and memory monitoring reflects a lack of the requisite abilities or even of skills. The child may simply not know where and when to use his abilities and skills.

Organization

Another set of memory-training studies assesses the child's ability to organize material to facilitate subsequent recall. Moely and her colleagues (Moely et al., 1969) have shown that preschoolers do not spontaneously organize material that they are asked to remember. Failure to organize the material during a study period correlates with poor recall scores. It could be that young children are unable to organize material into categories. Alternatively, it could be that they do not think of categorizing the material, in other words, that they do not make use of an ability they in fact possess. The latter proves to be the case. Given a brief period of instruction in the use of categories, the subjects in the above-mentioned study organized the material and used the organization in recall—as reflected in clustering scores and number of items recalled. We find this result to be of considerable interest. It suggests that the long-standing view of preschoolers as *unable* to classify by a consistent criterion is wrong. Further, it suggests a way to begin to study early classification abilities: by embedding the study of classification in a training paradigm that encourages the child to form categories.

Worden (1974, 1975, 1976) has done this type of study with children of elementary-school age. Worden was interested as much in her subjects' ability to organize input materials as in their ability to cluster their output on recall trials. She used a procedure that is standard in adult studies of such issues. This procedure involved allowing her subjects to sort verbal materials until they could do so consistently for two trials (compare Mandler and Pearlstone, 1966). A major result of such studies is that supposed developmental differences in the ability to impose structure on material and to use this structure in later recall are minimized under conditions that allow children to reach a stable sorting. Under such conditions, even 6-year-olds were able to sort materials according to taxonomic categories (Worden, 1974).

Several years ago Nash and Gelman undertook a study of preschoolers' classification ability that was motivated by considerations similar to those that motivated Worden's study. Preschoolers have been thought to lack the ability to classify, that is, to group materials according to meaningful or consistently applied taxonomic criteria. When tested for recall, preschoolers have not tended to recall material in meaningful classes or clusters, a result that has been thought to reflect their inability to categorize. Nash and Gelman did not consider the then-standard testing procedures likely to induce the preschoolers to use whatever classifying ability they might possess. They thought that training in which the children repeatedly sorted the same set of materials under the instruction "put together the things that belong together" would help induce the children to classify.

The study included three groups in order to assess the effect of different amounts of training. The first group received experience in categorizing blocks and toys (the block and toy condition); the second, experience in categorizing blocks (the block only condition); the third, no experience in categorizing (the no training condition).

Children in the block and toy condition were first shown a set of 18 wooden blocks, which varied in size (large, small), shape (rectangle, triangle, cylinder), and color (black, gray, white)—a set of materials much like the standard Vygotsky blocks. They were asked to "put the blocks together that belong together." If a child did not understand, the experimenter showed the child a way of doing the task and then asked him to try again. If he still did not understand, she made him copy a model and then try one of his own. After a child sorted the blocks once, he was required to sort them again in a different way. It was assumed that having to sort the same materials in two different ways would help the children understand the instruction. Note that the blocks could be grouped according to color, shape, size, or any combinations of these features. The unbalanced variation of dimensions (two sizes, three shapes, and three colors) was intended to emphasize to the child that there was no fixed "correct" number of categories or way of doing things.

The block training complete, the experiment continued on the next day with the toy training. The experimenter placed five clear plastic shoe boxes on the table and told the child that they were going to play a game much like the one they had played the day before. She would give him toys and he was to name each toy and place it in one of the boxes, keeping those toys together that belonged together. She then took one toy at a time from a bag and gave it to the child. There were 25 toys, representing 5 categories (fruits, vehicles, kitchen furniture, flowers, animals). If the child could not name a toy, the experimenter said the name and asked the child to repeat it (he often did so spontaneously). Training continued until the child sorted the toys the same way on two consecutive trials. However, the experimenter stopped asking the child to name the objects after the first trial. When a stable sorting was achieved, the child was asked why he had put the things together as he had. Note that children were judged to have met the criterion as long as they sorted the toys in a consistent manner; they did not have to use the experimenter's categories. The toy training allowed children to impose their own subjective organizations, thereby avoiding the possibility that the child's own categories would interfere with ones imposed by the experimenter (compare Mandler and Pearlstone, 1966).

Children in the block only condition received the block training described above but not the toy training. When they returned on the second day, they were shown each of the toys and required to name them; then they were allowed to play with them for approximately 15 minutes. Children in the no training condition received no categorization experience at all. They did, however, have an opportunity to play with both sets of materials (one on each of two consecutive days) for a period of time comparable to that covered in the training situations (approximately 15 minutes). When they received the toys on the second day, they were asked to name them. Here, as in the other conditions, if a child failed to provide his own label, the experimenter gave the common name and asked him to repeat it.

Free-recall testing occurred immediately after each child's experience with the 25 toys. The child was simply asked to tell the experimenter the names of the toys with which he had been playing. When it seemed as if the child could recall no further items (when he had been silent for one minute), the experimenter prompted (primed) him with the name of a particular object, usually from a neglected category. Thus, measures of recall could be computed for items recalled with and without priming.

A total of 162 children of ages 3, 4, and 5 participated in the study. The subjects were black or white and represented a lower-middle-class to middle-class socioeconomic population. Equal numbers of children from each of the three age groups were assigned to the three main conditions, yielding an *N* of 18 children in each of the nine conditions.

Nash and Gelman used a variety of recall measures. The main results are shown in Table 2.1. In almost all cases, the more training, the better the clustering scores and the better the recall. Some differences between the block and toy and block only conditions were not significant, although both groups that received training scored reliably higher on recall than the group that received no training.

Table 2.1 shows an effect that might be considered peculiar: Even children who received no training were likely to cluster their answers by categories. A score of 0.00 would mean that there was no such tendency; a score of 1.00 would mean that children always recalled items

TABLE 2.1. Free-recall phase of the Nash and Gelman experiment.

Age group and training condition	Mean number of items recalled: Without priming[a]	Mean number of items recalled: With priming	Mean clustering score: Without priming[b]	Mean clustering score: With priming
3-year-olds				
Block and toy	6.83	10.06	0.74	0.73
Block only	5.00	8.56	0.69	0.61
No training	4.22	7.33	0.50	0.58
4-year-olds				
Block and toy	9.67	13.00	0.77	0.79
Block only	8.30	12.89	0.62	0.62
No training	7.67	12.22	0.48	0.57
5-year-olds				
Block and toy	11.33	15.61	0.83	0.83
Block only	9.28	15.00	0.76	0.74
No training	9.00	14.72	0.59	0.63

a. Total possible is 25.
b. Measure of tendency to recall related items in a block. The formula used was $r/N - c$, where r equals number of clustered pairs recalled from the same subjective category; N equals total number of items recalled, including multiple occurrences; and c equals number of subjective categories.

within a category one after the other. The children in the no training condition did not cluster as much as the other children, but they did cluster.

This clustering in the recall performance of untrained controls is perhaps the strongest evidence that the training induced the used of a preexisting ability to categorize rather than instilled the ability. This interpretation is strengthened by an analysis of what the untrained children did while playing with the toys whose names they were eventually asked to recall.

Table 2.2 shows the number of children in each group who sorted the toys by taxonomic category. Note that at least 66 percent of the children in each age group who received instructions to sort the toys (block and toy condition) made use of taxonomic categories. But also note that at least 50 percent of the children in each age group of the remaining two conditions did likewise. Preschool children with no training in classification spontaneously arranged their toys in intelligible categories while playing with them.

TABLE 2.2. Number of subjects who sorted toys by taxonomic category in the Nash and Gelman experiment.

		Condition		
Age group	*N*	Block and toy	Block only	No training
3-year-olds	18	12	10	9
4-year-olds	18	15	13	9
5-year-olds	18	16	16	15

This last finding is the most desirable kind of evidence for our interpretation of many successful training studies. It is direct, as opposed to inferential, evidence that an ability that supposedly was to be trained in fact was already present. Indeed, the convincing character of this last finding emphasizes by contrast the unconvincing and limited conclusions about preexisting abilities that one can draw from training studies.

Interpretation of Training Studies

We have not exhausted the list of studies in which training seemed to uncover, rather than instill, some capacity. Trabasso, Deutsch, and Gelman (1966) challenged the idea that preschool children have great difficulty with tasks that require detecting that the reward assignments given to two stimuli (such as red and blue) have been reversed. Simply telling children who had learned $B+$, $R-$ that they were about to start a new game enabled them to solve the $R+, B-$ problem very rapidly. In other words, the young children could readily detect a reversal in reward assignment, provided they were cued to expect one. The effects of other studies suggest that children do not lack the ability to attend to certain variables in the environment. Luria (1961) demonstrated that, in a given test situation, a 2- to 3-year-old child failed to attend to the background of a visual stimulus, focusing instead on the figure. But Luria was able to alter the child's preference by labeling the background in a meaningful way. Gibson (1969) provided ample evidence that researchers can shift preschoolers' attention by using a training session.

Some might conclude that we are Platonists: that we believe training always simply uncovers an already available capacity. But our claim is much more limited. We believe that some training studies clearly have uncovered the existence of the very ability that they set out to instill. The fact that it took a training study to bring the ability to light is reason enough to resist the conclusion that the young child's capacity is the same as that of the older child. The young child needs the training, while the older child does not. But saying that some difference in ability or skill exists between preschoolers and older children is quite different from saying that preschoolers lack either the capacity for the task in question or the skills needed to show that capacity.

We do not mean to argue that the preschooler is just like the older child. Rather, our review of training studies is meant to demonstrate that we may expect to find overlap between the abilities of younger and older children—if only we will look. We would be foolhardy to reach any stronger conclusions. After all, many training studies have failed. Indeed, reviewers of conservation-training studies have only recently accepted the conclusion that conservation ability can be taught or uncovered (Brainerd and Allen, 1971; Beilin, 1971). It used to be a commonplace that conservation training failed to affect the child's performance on conservation tests (Flavell, 1963; Kohlberg, 1968). Kohlberg (1968) suggested that this ineffectiveness of training was quite general and not limited to conservation. While quite a number of training studies have been successful, there is no denying that failures abound, especially with respect to Piagetian tasks involving concrete operations (such as conservation, transitive inference, and class inclusion). Indeed, considerable skepticism still exists about the possibility of training preschoolers to conserve, classify, seriate, communicate in a nonegocentric fashion, and the like. (See Inhelder and Sinclair's criticisms [1969] of Kohnstamm's training technique for inducing class-inclusion skills [1967].)

The ambiguous results of training studies raise two questions. First, what direct evidence do we have that preschoolers possess cognitive abilities that fail to show up on the standard tasks, the tasks that are easy for 7-year-olds? So far we have advanced our position by indirect inference. If preschoolers utterly lacked the capacity in question, then training studies designed to teach it should falter; in Piagetian terms, without some relevant scheme to which the child can relate the input, there can be little learning (compare Inhelder, Sinclair, and Bovet, 1974). On the other hand, if preschoolers possess some capacity in the domain in which they are being trained, then it is possible that training studies encourage the display of this capacity rather than instill the capacity. In some training studies young children move rapidly from showing no evidence of the capacity or skill in question to showing strong evidence of it. We take this result as reason to accept the hypothesis that young children have more cognitive capacity than meets the eye. Since it is not possible to make this case for all training studies, it is necessary to provide *direct* evidence of preexisting abilities and skills. We must search for tasks that reveal the capacities or skills without any training.

Second, does not the mixed success rate of training studies mean that preschoolers cannot possibly be as able as we suggest? One might argue that if our hypothesis were correct it would always be possible to reveal hidden capacity through training. But our claim is not such a strong one. The many difficulties encountered in efforts to train various capacities suggest to us that differences in capacity between preschoolers and older children do indeed exist. Our position is that we will be better able to characterize these differences when we know more about what young children can do. The idea is to look for ways to compare, as well as to contrast, the preschooler with his older sibling.

Chapter Three

More Capacity Than Meets the Eye: Direct Evidence

On the one hand we argue that younger and older children have more in common than is generally thought. On the other hand we believe that the two groups must differ in ways that are not trivial. Same but different? We think so, and we are not alone. Recent work has begun to swing the pendulum to the view that at least some cognitive capacities of preschoolers are like those of adults.

Two Theories

Developmental psycholinguists point to the fact that preschoolers have remarkably good facility with the language of their community —a facility that suggests a very complex set of cognitive processes. Although questions remain about how complex or complete the 4-year-old's linguistic capacity really is (see, for example, C. Chomsky, 1969), there is no denying that the level of competence reached by this age is complex. Linguists and developmentalists alike have accepted the premise that such competence requires the postulation of an elaborate formal model—one that involves the ability to classify parts of speech and makes use of transformational rules like deletion, substitution, permutation, and so forth. It is hardly surprising to find a psycholinguist suggesting that similarly complex abilities exist in other cognitive domains. Indeed, one of the earliest suggestions that very young children share some common cognitive capacities with older children and adults came from researchers identified with the psycholinguistic tradition, Mehler and Bever (1967). More recently, Fodor (1972) has taken the view that the cognitive capacities of the preschooler are much like those of older children and adults.

Fodor's position derives from a consideration of Vygotsky's findings (1962) on the development of the ability to classify materials with respect to a criterion that is best described as the intersection of sets.

Young children fail the Vygotsky classification task, which requires them to sort blocks that vary in shape, color, height, and width into mutually exclusive subsets of tall and wide, tall and narrow, short and wide, and short and narrow blocks. Formally speaking, the subset of blocks that are tall and wide is said to be the intersection of the set of tall blocks and the set of wide blocks. At age 10 or 12, subjects readily discern the required solution. Fodor questions the use of such data as evidence for the position that preschoolers are unable to form concepts. He points out that such a position involves assuming that the requisite programs for concept formation are most appropriately described in terms of Boolean algebra, which is a formalization of the logic of statements that use the words *and, or,* and *not.* Fodor objects to the conclusion that preschoolers cannot form concepts because they lack the Boolean cognitive processes; he points out that it is quite unlikely that most adult concepts are so characterized (see, for example, Markman and Siebert, 1976).

Fodor asserts that evidence of very complicated cognitive processes in young children can be found "if you look in the right places" (1972, p. 93). His right places seem to be very young babies' categorical perception of phonemes (Eimas, 1974); babies' ability to detect perceptual invariants (Bower, 1971); 2-year-olds' ability to talk (for example, Brown, 1973); and babies' early recognition of faces (Gibson, 1969). Whether these abilities belong to the same domain as the abilities to conserve, monitor one's capabilities, construct representations of someone else's perspective, and so on surely can and will be challenged. We ourselves see problems in this regard. But for now let us accept Fodor's evidence on Fodor's grounds and consider the interpretation he offers.

Fodor grants the young child capacities that others would deny him. He views the young child as a "bundle of relatively special-purpose computational systems which are formally analogous to those involved in adult cognition" (1972, p. 93). The difference between the child and the adult is one of quantity rather than quality. The young child's use of the computational systems is tied to very specific situations—presumably because these systems were evolved (designed?) for species-specific purposes. The adult's use of these same systems is not so restricted. Cognitive development proceeds along two lines: the maturation of the processes involved in the computational systems and the extension of the operation of these computational systems to more and more domains.

Fodor does not provide a specific example of how development might proceed through the application of available computing programs to new problem domains. But Rozin (1976), who has developed a very similar argument, does. Lest we give the impression that one needs to be a philosopher-linguist to arrive at such a view, it should be noted that Rozin is a comparative psychologist. Indeed, until recently most of his work has been with animals. This work has led him to a theory of the phylogenetic development of intelligence that applies also to ontogenetic development, development within the individual.

Rozin begins by calling attention to the highly "intelligent" nature of many special-purpose behavioral mechanisms in animals. Foraging bees, for example, record the location of food sources in polar coordinates, with the home nest as the origin of the coordinate system and the sun as the point of angular reference. When they return to the hive they communicate the polar coordinates of the food source to their fellows by means of a dance upon a vertical wall of the hive. In the dance, the direction of gravitational pull serves as a substitute reference point for the angular coordinate. The angle to be pursued with respect to the sun is indicated by the angle of the dance with respect to the force of gravity. The bees have enough knowledge of celestial mechanics built into their nervous systems that, with the aid of an internal clock, they automatically adjust the angular coordinate to offset the change in the position of the sun through the course of the day! In other words, bees routinely make use of navigational techniques involving chronometric computation of the sun's course. Navigation by such "scientific" principles has traditionally been regarded as one of the triumphs of man's intellect. Any scout leader will testify that it takes more than a day to teach the average boy scout how to proceed from camp to some arbitrary point, specified in polar coordinates (compass angle and distance), using the sun as his point of angular reference. Yet such sun-compass navigation is common in bees and several other insects, as well as in some birds and some fish.

Rozin's point is that such "intelligent" behavior is founded as much upon genetically specified computational machinery as upon the animal's plastic capacity to benefit from experience. More accurately, an animal's plastic abilities (such as the bee's ability to learn the location of a food source), do not reflect some general-purpose faculty of association. Rather, they usually appear within genetically constrained behavioral circumstances. Rozin emphasizes that the genetically constrained behaviors evolved to serve specific rather than general purposes. It is highly unlikely that the foraging bee's navigational attainments derive from a general knowledge of celestial mechanics and the use of polar coordinate systems. Indeed, the "knowledge" of celestial mechanics that is implicit in the computations the bee must make in finding its way to and from food sources may be completely unavailable for use in other aspects of the bee's behavior. According to Rozin, this behavioral machinery may be inaccessible for use in contexts other than the specific context that shaped the evolution of the requisite neural machinery in the first place.

Rozin's thesis is that the evolution of general-purpose intelligence in higher mammals has involved the evolution of more general access to computational processes that originally served specialized behavioral purposes. He stresses the fact that even in humans there are many computational routines whose outputs are not generally accessible. For example, our visual system makes extensive computations that draw upon a great deal of implicit knowledge of geometric optics and trigonometry. The end result of the computations—our perception of the world about us—is generally accessible. But the intermediate stages in the computations are completely inaccessible. Indeed, even the raw sensory information upon which the computations are based is completely inaccessible. People are incapable of judging the size of the image projected from an object onto the retina (Hochberg, 1964) or even the part of the retina upon which that image is falling (Brindley, and Merton, 1960). Suppose people had to learn to say "Feed me, please," when and only when their retinal image of a refrigerator was larger than two millimeters. They would never master this simple task (or they would master it with great difficulty), because their learning mechanisms have little or no access to the necessary sensory information. On the other hand, learning to say, "Feed me, please" only when one sees a refrigerator more than four feet tall is easy, because our perceptual system, which provides no generally accessible signal indicating the size of the retinal image of an object, does provide a generally accessible signal indicating the size of the object itself. And this despite the fact that the signal indicating the object's true size derives from a complex computation that utilizes, among other things, the size of the retinal image.

Rozin emphasizes that we lack conscious (or generally utilizable) access not only to such things as the size of our retinal images but also, and more importantly, to many inferential principles that we routinely apply to certain restricted problem domains. If we tell the reader that the retinal image of x is 4.0 millimeters in diameter, that x is 20.0 meters from the lens of the eye, and that the lens is 1.8 centimeters from the retina, the reader may have some trouble recalling the trigometric principles by which to compute the size of *x*. Yet the reader's visual system makes routine use of those principles or an equivalent set of principles every waking minute in judging the sizes of objects in the world around him.

Rozin and Gleitman have applied this reasoning to the question of why some children have so much difficulty learning to read (Gleitman and Rozin, 1977; Rozin and Gleitman, 1977). They argue that learning to read requires gaining conscious access to the phonetic representation of the verbal sound stream. They further contend that this phonetic representation of the sound stream is not readily accessible to the conscious mind of many children, even though these children obviously compute and utilize such a representation in both the comprehension and the production of speech. Humans appear to possess special-purpose, genetically specified neural machinery for computing phonetic representations of the speech that they hear (Eimas, 1974).[1] This phonetic representation is an intermediate stage in the computation of a semantic representation of what they hear. Just as in the visual system, many intermediate stages of the computations carried out by the auditory system are not generally accessible. Most people, for example, have no access to the absolute—as opposed to relative—frequency of pure tones, despite the fact that information about absolute frequency is contained in a very simple way in the early stages of the neural signals in the auditory system. (Using biofeedback equipment to provide people with conscious access to the signals in their auditory nerve could endow all of us with perfect pitch.)

A strong selection pressure favoring the genes that specify the neural machinery required to compute a phonetic representation of human utterances has existed ever since speech became a significant part of human social interaction. How strong this selection pressure is and how long it has acted are indicated by the fact that the necessary genes are very nearly universal, that is, present in all humans. A comparable selection pressure in favor of genes that make it possible for humans to compute a phonetic representation of what they see on a printed page has existed in Western societies only since the emergence of widespread literacy less than 100 years ago.

Rozin and Gleitman argue that our comprehension of printed material is mediated by a phonetic representation. This view of the read-

1. Some question exists about whether this machinery is either uniquely human (Marler, 1977) or uniquely suited to the perception of phonemes (Cutting, 1977).

ing process, which we believe to be essentially correct, is controversial (see Smith, 1971, for a diametrically opposed view). Whether or not one believes that the fluent reader computes phonetic representations in the normal course of reading, it is undeniably the case that every fluent reader is capable of doing so: Every fluent reader will give decidedly nonarbitrary pronunciations of words he has never seen before. These systematic features in the pronunciation of unfamiliar written words demonstrably derive from the spelling rules. Spelling rules relate written English to a phonetic representation of spoken English. Thus, it seems hard to deny that an important aspect of learning to read is learning to compute a phonetic representation of written material by using the lawful relations between spelling and pronunciation. Rozin points out that learning to compute such phonetic representations of visual inputs must be very difficult if one does not have, at the outset, conscious access to the phonetic representation of what one hears. There is abundant evidence that many children do not have access to the phonetic representation of their own speech. Some modern methods for teaching reading therefore involve training that is designed to give children conscious access to the phonetic representation of speech (Liberman et al., 1977). Other programs try to circumvent this problem, in the hope that children who are not hopelessly frustrated in the early stages of learning to read will eventually gain access to the phonetic representations of speech, either through experience or through maturation (Gleitman and Rozin, 1973a, 1973b; Rozin and Gleitman, 1977; Gleitman and Rozin, 1977).

This analysis of the source of children's difficulty in learning to read illustrates the Rozin-Fodor contention that cognitive development may involve the development of the ability to use certain computational routines or the outputs of such routines in a variety of domains. The basic machinery required to perform certain feats of reasoning may be at work in certain domains long before the system as a whole develops access to these special-purpose computers and permits the machinery to be put to effective use in other domains.

The Evidence

We do not quarrel with Fodor's suggestion that young children have unrecognized cognitive abilities. We would not have written this book if we did. What we worry about is the view that the underlying capacities of the preschooler differ little from those of older children. Our concern stems from a consideration of the data base with which

Fodor supports his conclusions. Fodor takes the very young child's skill at face perception, phoneme perception, and language production as evidence that the cognitive capacities of the young child are much like those of older children and adults. While many studies have made it clear that the skills in these domains are in some ways comparable across ages, it is nontheless reasonable to believe that a 4-month-old infant's classification of phonemes is not quite the same as an adult's (Cutting and Eimas, 1975). Likewise, none would deny that the preschooler is able to recognize faces and produce complex utterances. It even appears that 4-year-olds, like adults, adjust the syntactic complexity of their utterances to match the different receiving capacities of 2-year-old, peer, and adult listeners (Shatz and Gelman, 1973). But again, it is far from clear that the same underlying abilities enable both preschoolers and adults to recognize faces (Carey and Diamond, 1977). The syntactic and communication abilities of preschoolers also differ from those of adults in ways that may be qualitative as well as quantitative. Preschoolers can work with some syntactic deep structures but not others (C. Chomsky, 1969). And, although they can alter the length of their messages, it remains to be determined whether they honor the same rules of conversation as adults do (Gelman and Shatz, 1977). We must be careful not to jump to the conclusion that the undeniable existence of some common capacity and skill in the domains of phoneme perception, face perception, and language use means that young children and adults possess the same underlying cognitive structure.

We also question another step in Fodor's argument. Since preschoolers (and even infants), the argument goes, have the same capacities as adults in certain cognitive domains, it follows that the two groups have the same capacities in many cognitive domains. If it should turn out that phoneme perception in the infant is identical to that in the adult, surely most cognitive developmentalists would not think it reasonable to conclude from this that the child's reasoning and mnemonic processes also are identical to those of the adult. As we have pointed out, the commonplace observation is that preschoolers fail reasoning tests that older children pass with ease. Before it can be argued that the preschooler's capacities are like the adult's, it is necessary to produce evidence that preschoolers can accomplish tasks similar to the ones at which they so regularly fail.

What direct evidence do we have that preschoolers do in fact classify stimuli by consistent criteria, or conserve, or monitor their memory capacities, and so on? The evidence we have presented so far is mostly indirect, coming from training studies. Even if we are correct that this training evidence implies that preschoolers have more competence than meets the eye, we still must face the fact that preschoolers need training while older children do not. Why? Perhaps Fodor's ideas apply here: Perhaps training studies allow the child access to his underlying capacities. But before we so conclude, we need direct evidence that the child has such underlying capacities. We are not willing to take a stand until we can demonstrate this one way or the other.

Our hesitation here might be taken to reveal an unduly conservative stance. But we think not. More accurately, it reveals an uneasiness about Fodor's treatment of evidence that is inconsistent with his position. He dismisses the young child's failure on the classification task on the grounds that the task criterion is unreasonable. We agree that it is a bit odd to conclude that a young child lacks the ability to "form concepts" because he fails to classify blocks using a criterion that involves the intersection of two sets. The fact remains, however, that older children and adults do use such a criterion and do perform the task successfully. No matter how we push and pull, it remains clear that the young child is different in some way.

Nor is it only Vygotsky blocks or the intersection of sets that stymies the young child who is asked to perform a classification task. Bruner and his colleagues (Bruner et al., 1966) report on a series of classification tasks that yield developmental functions much like Vygotsky's. For the most part, Bruner's tasks do not involve intersecting sets. The child is mostly left on his own to group the objects as he likes and to give his own reasons for judging items to be alike. Whether the task comes from Bruner, Piaget, Werner, or Vygotsky, and no matter what the procedure, criterion, or theoretical bent, the same results occur over and over again: Preschoolers seem to be unable to apply consistent criteria when sorting a set of materials. Objects are grouped because they form nice pictures; because they are fun to play with; "just because"; and so on. It seems not to matter that some of the red and green, large and small, tall and thin objects are in one group, some in another, and still more in another. Maybe it shouldn't matter —but that is not the issue. The issue is why the older child and the adult typically assume that materials are to be sorted by consistent criteria. For them, all the reds go in one pile, all the greens in another, all the yellows in still another. If the young child is so much like the adult, why is his performance so different? Young children do behave as if each part of speech—noun, verb, adjective—is to be kept in a separate bin. Interpreting this behavior as evidence of a classification skill makes it all the more surprising that they fail to apply this skill in a trivial situation.

Rather than dismissing failure on a trivial task as uninteresting, we consider failure on a trivial task remarkable. Our curiosity about cognitive development derives to a large extent from observations of how tasks that are easy for us as adults are hard for young children. The prevailing view is that such discrepancies reflect an underlying lack of capacity in the children. We hesitate to accept this conclusion, for all the reasons outlined in the preceding chapters. But rather than ignore the discrepancies, why not try to demonstrate that the classifying abilities of younger and older children have much in common? More generally, why not try to demonstrate competence in young children in the cognitive domains where incompetence is assumed? The challenge is to produce compelling evidence that the preschooler is capable of treating quantity as invariant, of sorting materials according to consistent criteria, and so on. Fodor's intriguing views of development are all too readily undermined by the claim that it is difficult, if not impossible, to find the common ground between the ability to recognize faces and the ability to use arbitrary classification criteria, to conserve, or to adjust messages to the needs of a listener. Theories that emphasize the cognitive deficiencies of the preschooler focus on the latter abilities. Fodor needs evidence of tasks in *these* domains of cognition that show the preschooler to be like his elders.

The type of evidence that could support Fodor's ideas does exist. A number of recent studies of young children have demonstrated their ability to do things they have been thought incapable of doing. Lest the reader misunderstand the moral we wish to draw, we emphasize that each of these studies used procedures that differed from previously employed procedures. It must not be forgotten that when tested by the previously employed procedures preschoolers routinely fail while adults and older children routinely and easily pass. Therefore we cannot conclude from their success on the newer tests that preschoolers have exactly the same abilities as older children.

Classification

Discrimination Learning. As emphasized earlier, young children seem to be unable to use consistent criteria in sorting stimuli that vary in more than one way. Results from many discrimination-learning studies buttress this conclusion. Discrimination-learning tasks require the subject to learn to sort a set of stimuli according to attributes that the experimenter defines as correct. The stimuli are typically line drawings, drawings incorporating variation along several dimensions. To succeed at this task, the subject must use the experimenter's definition of what constitutes the correct (relevant) dimension: the one the experimenter reinforces. He must also ignore the irrelevant dimensions, the ones the experimenter does not reinforce. A host of studies document the tendency of young children to be distracted by the irrelevant attributes (for example, Bruner et al., 1966; Gelman, 1969; Gibson, 1969; Hagen and Hale, 1973; Maccoby, 1969). Some evidence even indicates that young children learn *more* about irrelevant attributes than do older children (see, for example, Crane and Ross, 1967; Gibson, 1969; Kemler, 1972). The presence of irrelevant attributes, then, seems to keep younger children from focusing on and using the experimentally defined relevant information. The fact that younger children are less inclined than older children to focus on relevant information (Gibson, 1969) goes a long way toward explaining young children's difficulty with discrimination tasks that involve several irrelevant dimensions (see, Adams and Shepp, 1975).

As reasonable as these explanations may seem, the results have always bothered us. Watching preschool and kindergarten children, we get the impression that they are successfully negotiating a complex world. In doing so, they must keep track of relevant information no matter how many irrelevant attributes surround it. We are reminded of an observation made by two of Gelman's former graduate students. Heidi Feldman and Susan Goldin-Meadow took a bag of toys to the home of a 2½-year-old child who was to participate in a memory study. Upon seeing the toys, which varied in color, shape, function, material, and so on, the child said, "I'll play with the red ones," and proceeded to pull out every red toy. The experimenter's subsequent request to "put together the ones that belong together" was greeted with a blank stare. Either the child did not understand the question or the task as presented by the experimenter was not a game he wanted to play. The child had just demonstrated spontaneously that he was capable of classifying materials, yet he responded to the experimental situation with behavior that has often been interpreted as evidence of an inability to classify.

This anecdote highlights the need to embed the experimental task in a game that preschoolers are likely to enjoy playing. Also, of course, the game should maximize the likelihood of the child understanding what the experimenter wants of him. Kemler's ingenious

(1972) twin game meets these standards. It fulfills the requirements of a standard discrimination-learning task, yet it has the ring of a game that would surely appeal to a young child.

Kemler presented her subjects with cards displaying line drawings of young girls. The same basic form (Figure 3.1) appeared on each card. In addition to the basic form, each card could contain one value for each of five attributes. The attributes were types of clothing: hat (crown or party hat), necklace (medallion or beads), hair ribbons (red or blue), belt (leather belt or sash), and glasses (regular glasses or sunglasses).

Before starting the experiment proper, Kemler introduced each child to the situation with a story about identical twins, Amy and Betty. Since they were identical twins, their classmates and teacher

Figure 3.1. Copy of a stimulus card used in Kemler's twin game (1972).

had trouble telling them apart. So their teacher bought the twins special clothes to make it easier to tell them apart. As the story proceeds, the new items of clothing are named and shown. The subject is told that the teacher bought two different hats, a party hat and a crown; two different belts, a leather belt and a sash; and so on. A cut-out of each item is shown. The subject is then told more about Amy and Betty. It seems that the twins decided to play tricks with their new clothes. So just as their friends were beginning to remember that Amy wore one kind of clothing and Betty another kind, they would switch clothes to confuse their friends. They were a little kinder to their teacher. They told the teacher that they had a secret; the secret was the one item of clothing that they promised not to switch all day. Every day there would be a new secret, which would be true for the whole day.

In the game the subject was to figure out the twins' secret, that is, to determine what one thing Amy and Betty would not switch throughout the school day.

In her work with kindergarten children, Kemler tested 48 children on each of 5 problems involving 4 attributes of clothing. The 5-year-old subjects solved 84 percent of their problems in fewer than 40 trials and 73 percent in fewer than 20 trials. As Kemler (1975) points out, this is a notable success rate for such young children on 4-attribute discrimination tasks. Compare the results of Adams and Shepp (1975), who report that 81 percent of their kindergarten subjects and 84 percent of their second-grade subjects failed to reach criterion within 20 trials on a standard discrimination-learning task involving but 2 dimensions. In Kemler's situation children of kindergarten age were able to find and use a consistent criterion. They also adopted new criteria as new problems were introduced. A third result was even more unusual: These children were able to verbalize the criteria they were using. We do not mean to suggest that Kemler's technique yielded no developmental functions. The 5-year-olds were less inclined than the older children to use stimulus-feedback information from earlier trials when selecting criteria on later trials. Still, Kemler's results go a long way toward challenging the view that 5-year-olds fail to use consistent criteria in sorting materials, are unable to use different criteria on the same set of materials, and are unable to verbalize their hypotheses. We cannot help but wonder what other cognitive talents lie ready to be discovered by experimenters who have the knack of designing tasks that captivate the child rather than bore or confuse.

We do not mean to suggest that failure to use game-like tasks is the major reason for researchers' negative findings about preschoolers' abilities. The point is simply that a game can help motivate the child and relate the task to the child's approach to problems. In Kemler's case, embodying the task in a detective game apparently made it easier for the child to understand what the experimenter wanted him to do.

Rosch's Work on Classification. If designing the task to suit the child is so important, perhaps children's performance on sorting tasks partly depends on what types of objects they are asked to sort. The work of Rosch and her collaborators suggests that children are more likely to use consistent criteria if the sorting task involves "natural" categories rather than the arbitrary categories typically used in such tasks.

It is easy to think of examples of arbitrary categories, such as the category "tall and narrow" in the Vygotsky task; but what are the natural categories? Rosch provides evidence that, of the many levels of abstraction humans are able to impose on objects, there exists a preferred *basic level* of abstraction (Rosch and Mervis, 1975). The preference for this basic level is due in part to the basic structure of the stimuli we encounter in the real world. Attributes such as feathers and wings do not occur in random combination; Rosch suggests that the basic level of cognitive categorization reflects such real-world correlations.

The way we interact with the world also helps determine the basic level of categorization. Given information-processing constraints, the basic level should be the one that provides the most information with the least cognitive effort. This requirement leads to a definition of the basic level of cognition: the level at which objects share the most attributes that are relevant to humans. Since humans interact with objects by consistent motor programs (chairs are to sit on, flowers are to pick, and so on), common motor programs are also likely to form part of the operational definition of the basic level of cognitive categories (Rosch et al., 1976).

When adult subjects are asked to list attributes and describe movement routines for objects that can be named at three levels (for example, city bus, bus, vehicle), they indeed behave as if there is a basic level for forming abstractions about objects. They list many attributes that basic-level objects have in common, but they can think of few, if any, attributes that superordinate objects have in common. For example, members of the basic-level category *chair* have seats, legs, and surfaces to sit on, but it is difficult to identify the properties that are shared by members of the superordinate category *furniture*. Basic-level objects share common movement schemes (chairs are sat on) whereas superordinate objects do not. We could sit on a television console or the dining room table, but we rarely do so.

Rosch's definition of a natural category is one that involves basic objects. Do children categorize such objects more readily than other objects? To find out, Rosch and her coworkers investigated the ability of children to sort basic-level objects into basic categories and into superordinate categories. Subjects were kindergarteners, first graders, third graders, and fifth graders. Stimulus materials were color photographs of objects that fit into superordinate and basic categories. The superordinate categories (with their basic categories in parenthesis) were as follows: clothing (shoes, socks, shirts, pants); furniture (tables, chairs, beds, dressers); people's faces (men, women, young girls, infants); and vehicles (cars, trains, motorcycles, airplanes). Subjects in the superordinate sorting condition were given one picture of each of the four objects in each of the four superordinate categories: one shoe, one sock, one shirt, one pair of pants, one table, and so on. Subjects in the basic sorting condition received four distinct pictures of one basic object from each of the four superordinate categories. Thus, for example, a subject in the basic condition might receive four shirts, four beds, four faces of infants, and four airplanes.

The results of the experiment are easy to report. Only half of the kindergarten and first-grade subjects used the superordinate criteria, while the older children and adults consistently used such criteria. However, the age groups showed no differences in the ability to use basic categories. Save for one kindergarten child and one first-grade child, all children sorted consistently according to basic category. A kindergarten child may not know that motorcycles and airplanes "belong together" as vehicles, but he does know that four different motorcycles belong together. Indeed, Rosch and Mervis suggest that 3-year-olds would do about as well (that is, nearly perfectly) on the basic-level sorting task. Using a simplified (oddity) sorting task, they were unable to detect any developmental trend from age 3 up to adulthood.

These results should not be taken to indicate that no interesting developmental differences exist. But they do call into question the widely held assumption that preschoolers are unable to sort stimuli according to consistent criteria. Furthermore, they draw attention to the need to consider the nature of the stimuli used in classification tasks. Some criteria for classification seem to be readily available to young children, while others do not. Rosch's work points to the need to determine why preschoolers have particular difficulties with superordinate criteria. (For a possible explanation, see Markman and Siebert, 1976.)

The Nonegocentric Preschooler

We have backed away from the idea that the use of a game-like task will automatically produce results showing the preschooler to be a competent being. Many of the studies that have been designed to determine whether preschoolers are able to communicate in a nonegocentric fashion (that is, to produce messages that are appropriate for their listeners) resemble games. Piaget explains something to a child and asks the child to repeat the explanation for another child. Despite the fact that this resembles a storytelling setting, the child mixes up the sequence of events and fails to honor causal connections. The product looks egocentric: It fails to provide the kind of information the listener needs.

The literature contains other cases in which children do poorly on a task even though they enjoy it. Glucksberg and his collaborators (Glucksberg, Krauss, and Weisberg, 1966; Krauss and Glucksberg, 1969) use the following technique for testing communication skills. Two children are placed at opposite ends of a table that has a partition in the middle so that the children cannot see each other. Each child is given a peg and a set of blocks with abstract forms printed on them. The game involves having one child describe a block in such a way that he and the other child can both place the same block on the peg at the same time. To succeed at the game the children must stack the blocks in the same order on both pegs. There is every indication that the children enjoy the game. And it can hardly be described as a "test" that requires the children to sit still and answer questions. Yet the 5-year-old children do not do well at the task. Even when shown the outcome of their efforts to stack the blocks in the same order, they fail. Why? Their descriptions of the blocks are often interpreted as indications that children of this age are egocentric, that is, unable to provide messages that are appropriate for a given setting. The child who says a block looks like "Mommy's shirt" does indeed seem egocentric.

But is the child egocentric? Not necessarily. Consider the task demands. The objects to be described are abstract forms; indeed, they have come to be known as *squiggles* (see Figure 3.2 for examples). The

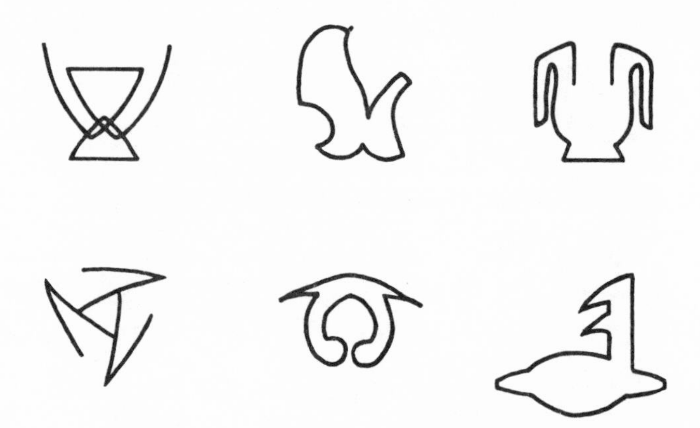

Figure 3.2. Examples of the "squiggles" used in Glucksberg, Krauss, and Weisberg's communication task. (From Glucksberg, Krauss, and Weisberg, 1966.)

child has to assign to each shape a label that uniquely identifies that shape. To do this, he has to identify the distinctive features of each shape. We know that children of this age have trouble with tasks that require recognizing distinctive features (Gibson, 1969). Thus the children's failure at the Glucksberg, Krauss, and Weisberg task may tell us little about their communication skills; they may be unable to represent the squiggles to themselves in terms of distinctive features. The task confounds the assessment of labeling ability with the assessment of communication skill.

A similar criticism can be made of Piaget's explanation task, in which children are asked to reproduce explanations about the working of water taps and syringes. Piaget's own work suggests that young children lack the ability to trace causality in physical events, or at any rate to talk coherently about causality. What does it mean if they fail to reproduce the causal sequence of the original account? Perhaps they did not understand the experimenter's account. Even if they did understand, they might have difficulty retrieving the explanation from memory (A. L. Brown, 1975b). We all look egocentric when we try to explain something we do not quite understand or cannot remember exactly. Readers who have less than a clear grasp of Einstein's special theory of relativity are invited to demonstrate this point for themselves by explaining the theory to a passerby. Chances are that your explanation will not make the theory clear to your listener.

Does this mean that you are fundamentally egocentric? Of course not. Whenever possible we avoid explaining what we do not understand, precisely because we know that such an explanation will violate the rules of conversation, rules that say "select appropriate messages, be clear, be precise," and so forth (Grice, 1975).

The point is made. The child may seem egocentric because he lacks knowledge or memory skills rather than because he lacks an intent to communicate as well as possible. How can we control for this possible confounding? First, ask the child to talk about things he is likely to know about. Second, choose to measure a response that he is likely to have available; avoid requiring responses that he is unlikely to have available. Shatz and Gelman (1973) followed these rules when they asked 4-year-old children to explain the workings of a toy. They measured the child's communication skill in terms of his tendency to adjust the complexity of his syntax when talking to listeners of different ages. Toys are an obvious choice. Why syntactic measures? Because 4-year-olds have a remarkably rich repertoire of syntactic constructions. Therefore it is *possible* for them to select different levels of syntax to match the needs of different listeners. For example, they might use shorter and simpler sentences, when talking to younger children.

When 4-year-olds talk to 2-year-olds about the workings of a toy, they do in fact tend to use short, simple utterances. With peers and adults they use longer and more complex utterances. This is true whether the 4-year-old is talking to his sibling or not, whether he has a sibling or not, and whether he is male or female. When the child is allowed to choose the topic of conversation, much the same result holds. And, besides being shorter and less complex, the utterances directed at the 2-year-old serve a different function and contain somewhat different messages (Gelman and Shatz, 1977). Speech to 2-year-olds serves to "show and tell": to direct, focus, and monitor attention. Speech to adults, on the other hand, includes talk about the child's own thoughts, seeks information, clarification, or support from the adults, and marks the child's recognition that he may not be correct about his assertions. The adult is treated as one who is in a position to offer advice, to challenge, and the like. Clearly the child selects his messages to match the needs and capacities of his listeners.

This work on the young child's conversations leads us to the conclusion that preschoolers are not always egocentric communicators. Further, it implies that the young child has formed representations of others—representations that involve such variables as cognitive capacity, age, linguistic level, and attention span. These representations appear to help guide the child's choice of responses.

Some readers may protest that we grant too much to the preschooler. After all, our subjects talked in the presence of their listeners, who may have provided subtle feedback. And it is possible that we have uncovered one more of the special and limited abilities programmed into the human species. Human language does have an elaborate apparatus for producing sentences of different types to meet different communication needs (Sachs and Devin, 1976). To answer this protest we need a way to demonstrate both that preschoolers can represent other individuals when those individuals are not present and that such representations control judgments and selections other than linguistic ones.

Shatz (1973) addressed these issues by taking advantage of young children's familiarity with birthday parties. Anyone who has attended such an event is aware of the role that the giving and receiving of presents plays in the life of the preschooler. Shatz designed a task that required preschoolers to choose presents for children of different ages. She assigned her subjects—28 4-year-olds and 30 5-year-olds—to one of two conditions. Those in the 2-year-old condition were asked to select a toy for a 2-year-old and then one for themselves. Those in the peer condition were asked to select a toy for a 4-year-old and then one for themselves. Depending on which condition they were in, they were shown a picture of either a 2-year-old or a 4-year-old boy they did not know. The use of pictures prevented the children from basing their choices on feedback from the persons for whom they were choosing.

The experimental stimuli consisted of four toys: two that were deemed appropriate for 2-year-olds (a pull-toy and a set of stacking cups); and two that were deemed appropriate for 4-year-olds (a set of sewing cards and a magnetic board with letters and numbers). The experimenter showed each subject a picture and asked the subject to help choose a birthday present for the child in the picture. After the child was shown the toys, picked one for the birthday child, and answered the experimenter's questions, he was told to pick a toy that he wanted to play with.

The results of Shatz's experiment were as follows: Subjects who were choosing gifts for 2-year-olds had a significant tendency to pick toys that were appropriate for 2-year-olds. Subjects in the peer condition likewise picked appropriate toys for the 4-year-old in the picture. Similarly, children who chose a present for a peer were more likely to pick the same toy for themselves than were children who chose a present for a younger child. The children rarely justified their choices with egocentric explanations such as "I like it." Instead they referred to the cognitive or affective characteristics of the receiver. Subjects who chose gifts for the 2-year-old explained, "I didn't pick this (the number-letter board) because he can't read," or "he'll like the noise (of the pull-toy)." As in the Shatz and Gelman study, the skills of the 2-year-old recipient seemed to be relevant factors in determining responses.

Markman (1973) provides support for Shatz's conclusion that preschoolers can accurately represent the needs and capabilities of individuals who are not physically present. Markman had 5-year-olds make predictions about their own capacities to remember and to perform motor tasks. The memory-monitoring task, adapted from Flavell, Friedrichs, and Hoyt (1970), required children to indicate whether or not they would be able to remember a given number of pictures. At first they were shown 1 picture and asked if they could remember it; then 2 pictures; and so on until 10 pictures had been shown. After predicting their own performances, the children were asked to predict how well a 2-year-old and a teenager would do on the same task. The motor tasks were similar in form. One involved predicting how far the subject could jump, then predicting how far a 2-year-old and a teenager could jump. The other involved predicting the number of small rubber balls the subject, a 2-year-old, and a teenager could pick up at one time.

Markman's results contain a wealth of interesting detail. For one thing, predictions about the subjects' own abilities tended to fall between the predictions about 2-year-olds and those about teenagers. It is of particular interest that the kindergarten children made differential predictions about the memory and motor skills of 2-year-olds. Most of them considered 2-year-olds to be able to handle the motor tasks at some level: to jump a little way and pick up one or two balls. In contrast, most of them predicted that the 2-year-olds would not remember anything. Kindergarten children are unlikely to have read developmental texts on the normative capacities of 2-year-olds; yet they correctly judge 2-year-olds to be more advanced motorically than cognitively.

Shatz's and Markman's results make it hard to maintain the view that preschoolers are unable to represent the capacities of others or to use these representations when making judgments. Moreover, the display of such capacities does not seem to be dependent on the physical presence of the other individuals.

Given the findings of Gelman, Shatz, and Markman, we are hardly surprised to read reports that preschoolers do well on visual perspective tasks (Lempers, Flavell, and Flavell, 1977). In one such task, children ranging in age from 1 to 3 were handed, one at a time, photographs of familiar objects. The face side of the photograph was always toward the child; however, some photographs were right-side-up and others upside-down. The child was asked to show the pictures to an observer (usually his mother) who was seated across from and facing him. This task was designed to test the children's egocentricity. An egocentric child might be expected to show the observer the back of the picture; a nonegocentric child should make some effort to turn the picture toward the observer, thereby depriving himself of a view of it. Of the 12 2-year-olds tested, 9 turned the right-side-up pictures away from themselves and presented them vertically to the observer. With the upside-down pictures, 10 of the 12 2-year-olds did likewise, although they did not necessarily show them right-side-up. All 2½-year-olds and 3-year-olds reoriented the pictures to face the observers, and most placed them right side up. Only the very youngest children showed any sign of being unwilling to turn the back of the picture toward themselves. Even these children, however, did not adopt the obviously egocentric strategy of continuing to look at the picture themselves and showing the back to the observer. Instead they tended to place the picture horizontally between themselves and the observer so that both could see it. Of the 12 1-year-olds in the study, 8 adopted this procedure, leading the researchers to suggest that "the very young children may not yet be able to deprive themselves of the percept while showing others" (Lempers, Flavell, and Flavell, 1977, p. 20). Not even the youngest children can be called egocentric, however, since they did produce a percept that someone else could look at.

Fodor's argument, that there is much capacity to be uncovered in preschoolers if only we look, sounds more and more credible. Shatz and Gelman started with a linguistic response system, a system that seemed particularly suited to revealing communication skill. Their initial results led them to postulate even greater capacities than they had expected to find. This in turn led Gelman's students to search for—and to find—corroborative data. The work on classification and visual perspective-taking shows that communication skills are not the preschooler's only "hidden talents." We cannot help wondering what other rich veins of cognitive competence lie waiting to be discovered in young children.

Chapter Four

Number Concepts in the Preschooler?

Preschoolers fail the Piagetian number-conservation test. Many researchers have concluded from this failure that preschoolers lack most of the principles necessary to a mature conception of number. Piaget (1952) has asserted that young children judge the numerosity of an array of items by attributes such as length or density rather than by number. Also, the child's reasoning about number has been assumed to reflect the inappropriate attributes from which he judges numerosity. For example, the child has been said to lack a number-invariance principle: to believe that spreading out or contracting an array of items changes the number of items in that array.

Mehler and Bever were among the first to argue that young children were not so incompetent in the number domain as had been supposed. A review of their work and the controversy it aroused provides a springboard for our presentation of research by Gelman and her students on young children's number concepts. In fairness to Fodor we should correct the impression we gave in Chapter 3 that he ignored the question of whether evidence existed of early capacity in domains in which preschoolers had typically been considered incompetent. Fodor did cite work by Mehler and his collaborators (Mehler and Bever, 1967; Bever, Mehler, and Epstein, 1968). Since this work has been challenged on several fronts, however, we deferred discussion of it until after our general survey of the evidence in support of our belief that young children have more capacity than meets the eye.

Mehler and Bever (1967) had children make judgments about the relative numerosity of two rows of objects. To start, the children were asked whether two rows of equal number (four items) and equal length contained the same number of items. Then one of the rows was modified so that it contained six items but was shorter than the unmodified row of four items. Each child was tested with pieces of candy and with clay balls. In the condition with pieces of candy the child was asked to choose the row he wanted; in the condition with clay balls he was asked to indicate which row had more. Mehler and Bever reasoned that if the children chose the six-item array despite the fact that it had been made shorter than the four-item array, they would be demonstrating an ability to conserve number. As it turned out, children of age 2 years, 6 months to 3 years, 2 months responded correctly, but those of age 3 years, 2 months to 4 years, 6 months did not. Still older children, over age 4 years, 6 months, responded like the 2½-year-olds.

These results led Mehler and Bever to advance an interesting model of cognitive development. They grant the very young child cognitive capacities much like those of older children (and presumably adults). They even suggest that the same formal models that describe advanced cognitive capacities hold for the cognitive systems of the 2-year-old. "The fact that the very young child successfully solves the conservation problem shows that he does have the capacities which depend on the logical structure of the cognitive operation" (1967, p. 142). Why then do children between the ages of 3½ and 4½ fail to use this capacity? Mehler and Bever answer that part of cognitive development involves the acquisition or generalization of perceptual strategies. As these come into play they may interfere with the existing cognitive capacities—so much that the child may give wrong answers even though he has mastered the processes that should lead to the right answers. The child must then learn to resist the pull of competing strategies or to recognize the conditions under which various strategies should be used. Bever cites studies of sentence comprehension as further support for this view of development (Bever, 1970).

Mehler and Bever's data and their interpretation of their findings have been challenged by a number of researchers. The reliability of their results has been questioned. Hunt (1975) shows that the results are very susceptible to experimenter bias: Telling an experimenter that 2-year-olds will do well at the task makes that experimenter more likely to replicate the original finding that the 2-year-old is a conserver. Even if the findings are reliable, it is not clear whether children as young as 2½ years understand the term *more* (see, for example, Beilin, 1968, 1975; Donaldson and Balfour, 1968; Palermo, 1973). The choice of a conservation test is another subject of dispute. Beilin (1968) and Piaget (1968) both argue that the issue of whether the original data reveal an early ability to conserve number is moot because the test procedure used was not the standard conservation test. Mehler and Bever are seen to have made two procedural errors: They failed to test the children's ability to conserve equality, and they confounded the transformations of displacement and addition. These mistakes make their results ambiguous about whether very young children know that the number of items in a set remains the same despite irrelevant transformations that are performed on the set.

Piaget argues that these results reveal the nature of the preschooler's quantification strategies but not a number-invariance scheme. According to his interpretation of the results, very young children judge the relative numerosity of arrays on the basis of their density; slightly older children rely on length cues; and still older children use a principle of one-to-one correspondence. He points out that Mehler and Bever failed to control for the possibility that the youngest children could base their choices on density. Adding items to a row while also shortening the row gives the children two reasons —density and number—to say that row "has more." In fact, Piaget says, very young children perceive the shorter and more numerous row as having more because it is denser, not because it is more numerous. Piaget also finds the errors of children in the second age group easy to explain: They choose the longer row because they have switched from density to length as a basis for judging quantity. Since length and numerosity varied in opposite directions in the Mehler and Bever experiments, these responses showed no ambiguity. Finally, Piaget argues, still older children rely on a principle of one-to-one correspondence and give correct answers for the correct reason; that is, they base their judgments on number. Since Piaget considers the ability to apply the one-to-one principle prerequisite for the ability to conserve, he does not credit the 2-year-old with mastery of number conservation.

Piaget's counterinterpretation of Mehler and Bever's results is an example of the scrutiny applied to studies that claim to find cognitive similarities between preschoolers and older children. It is not enough that both groups give the same answers; they must be shown to use the same processes to arrive at those answers. Any ambiguity about processes, especially if the test procedure can be faulted, is resolved in favor of differences between preschoolers and older children. Why? Partly, we believe, because of theoretical biases. Is it not possible that researchers are so molded by the belief that the young child is cognitively inept that they choose to interpret away any evidence to the contrary? Witness the plethora of follow-up research designs that seek to support the position that the young child does not know about quantitative invariance (for example, Beilin, 1968; Rothenberg, 1969).

We could go through the exercise of pointing out the various ways in which our evidence that preschoolers possess classification capacities and nonegocentric communicative capacities is likewise "inadequate." The evidence of competence in these domains does not prove that preschoolers actually possess the cognitive structures that Piagetian tasks are designed to detect. But we see no point in carrying through such a critique. As we argued in Chapter 1, we do not accept the view that a given capacity is tapped by one and only one task. Furthermore, we detect a tendency to reify Piagetian theory in the criticism of experimental results that appear inconsistent with the theory. The young child cannot conserve number, it is said, unless he understands the principle of one-to-one correspondence. It is also said that the young child judges number on the basis of density. If so, he must lack the principle of one-to-one correspondence, and therefore, the argument continues, he cannot treat number as invariant under spatial dispersions. Such an argument reifies the hypotheses that one-to-one correspondence is *the* cognitive basis for numerical equivalence judgments and that very young children base their numerical judgments on density.

We prefer another type of response to Mehler and Bever's results and the theoretical questions they raise. Namely, we prefer to search for tasks that unambiguously determine whether or not the young child honors number-invariance rules. We know that it is possible to find such tasks (see Gelman, 1972a). Once they are found, we can proceed to determine the extent to which the young child shares cognitive processes with older children. In other words, why not take Mehler and Bever's conclusions seriously enough to test them? Likewise, we might attempt to test the counterexplanations. Do very young children judge number on the basis of density? If they do not, then whatever the methodological pitfalls of Mehler and Bever's procedure, their results cannot be explained away by an appeal to judgments based on density. As it turns out, preschoolers do *not* use density cues in comparing the numerosity of sets (Gelman, 1972a; Lawson, Baron, and Siegel, 1974; Smither, Smiley, and Rees, 1974).

The ambiguity of Mehler and Bever's results casts doubt on one source of critical evidence in favor of Fodor's theory. We need more compelling evidence that the preschooler shares some of the adult's notions about number. Finding such evidence has been the goal of Gelman's research. This quest was undertaken not as a defense of

Fodor but rather as a consequence of Gelman's belief that the young child must be afforded every opportunity to reveal what cognitive principles he does have. When the opportunity is provided, the story of the preschooler's numerical capacities turns out to be lengthy and complicated. Likewise, for the *what* and *how* questions of cognitive development.

Chapter Five

What Numerosities Can the Young Child Represent?

The teaching of mathematics in general, and the teaching of arithmetic in particular, embody a tension between emphasizing techniques for obtaining a correct numerical answer to a problem and emphasizing the principles that underlie the manipulation of the abstract mental entities called numbers. The "practical" school focuses on teaching a child to add 13 apples to 11 apples and get 24 apples. The "theoretical" school focuses on teaching the child to recognize that addition is commutative: that 13 apples plus 11 apples yields the same number of apples as 11 apples plus 13 apples. In the practical school, a child whose homework paper says that $11 + 13 = 24$ and that $13 + 11 = 14$ is faulted for applying the techniques of addition incorrectly in the second problem. In the theoretical school, the child is faulted for failing to realize that the two problems should yield the same sum, never mind what that sum is. If the child gave 14 as the answer to both problems, the fault would be heightened in the eyes of the practical school and quite possibly diminished in the eyes of the theoretical school.

The two schools are focusing on two different aspects of the child's arithmetic capabilities. One focuses on the ability to obtain correct numerical representations of sets. The other focuses on the ability to reason about number and to recognize abstract properties in the operation of addition. The fact that $(a + b)$ equals $(b + a)$ is an abstract property of the addition operation, technically known as the commutative property, which holds regardless of the actual numerosities represented by the algebraic symbols a and b.

From the standpoint of arithmetic as a practical aid in everyday life, it is more important to be able to compute accurately the sums $11 + 13$ and $13 + 11$ than to realize that from the abstract point of view these sums must be the same. From the standpoint of arithmetic as the cornerstone of higher mathematics, the opposite is true. A mathematician who failed to correctly add 11 and 13 would inspire only indulgent chuckles among his colleagues; a mathematician who did not realize that addition was commutative would be drummed out of the profession.

Developmental psychologists, however, must analyze both of these aspects in order to gain insight into the child's ability to reason about numerosities. Although professional mathematicians are more than willing to reason about number algebraically, that is, without any actual numbers, young children are not. When the young child encounters numerosities that he can represent numerically, he can bring to bear reasoning principles of surprising sophistication. But these reasoning principles do not come into play in the absence of specific numerosities. Furthermore, the numerosities (sets of one or more objects) must be small enough for the child to determine the number that represents them. A set of 57 toy mice, although indisputably a specific numerosity, does not lead to numerical reasoning in the 3-year-old child. The child lacks the ability to determine how many items there are in such large sets. When asked to reason about large sets, he does poorly—to say the least. In contrast, a set of 3 toy mice can and does lead to numerical reasoning in the 3-year-old; the child of this age is able to obtain an accurate numerical representation of small sets (Gelman, 1972a). It is difficult to evaluate the young child's numerical reasoning without understanding when and how the child obtains numerical representations of real-world numerosities. These representations are the grist for the reasoning mill. We must also understand the techniques the child uses to compute the numerical representations of the outcomes of his reasoning. His reasoning principles may tell him that only subtraction can reduce numerosity, but they do not tell him how to compute a numerical representation of the set that must be subtracted from a set of four items to make it a set of only two items.

Number Abstraction versus Numerical Reasoning

A central part of our treatment of the development of numerical abilities in young children is the distinction between the processes by which the child *abstracts* number from an array and the processes by which the child *reasons* about number. In drawing attention to this distinction, Gelman (1972a) introduced the terms *estimators* and *operators*. Estimators are the processes that can be used to obtain a numerical representation of a particular array. Operators are the processes that define the outcomes of manipulating sets in various ways. The adult's operator principles specify that the transformations of adding and subtracting change number and that the transformations of item substitution, rearrangement, and displacement do not change number. The child who correctly counts the items in a set is using an estimator process. On the other hand, watching five rabbits go into a hat and then seeing that only two come out of the hat results in surprise because of the mediation of a cognitive operator. The operator specifies that placing arrays in hats and taking them out again is not a transformation that alters number. Similarly, the ability to predict that adding an object to a set will increase the set's numerosity is mediated by an addition operator.

The distinction between number abstraction and number reasoning processes is much the same as Gelman's earlier estimator-operator distinction. One reason for our shift in terminology is to avoid the use of terms that carry surplus meaning that we do not wish to incorporate into our working concepts. To many researchers, the word *estimate* means to reach an approximate, as opposed to an exact, answer to a quantification task. Klahr and Wallace (1973) use *estimate* in this way to distinguish between processes such as counting, which yield an accurate numerical representation of any array, and processes that yield only approximations. We now use the term *number abstractors* for processes that render either exact or approximate representations of the numerosity of given arrays.

Our other change in terminology, the substitution of *reasoning principles* for *operators,* is prompted by our wish to expand the domain of inferences about number. Inferences that involve operators (reasoning about the effects of manipulations) are an important subset of the numerical inferences people commonly make, but they are not the entire set. Other types of inferences concern the relations that hold between sets: the relations of equivalence, nonequivalence, and greater than or less than. And there are superordinate principles concerning the effects of a combination of operations. As adults, we know not only that addition increases set size but also that the effect of addition can be canceled by subtraction. In other words, we know the superordinate principle that subtraction reverses addition and vice versa. We refer to this kind of principle as a higher-order principle of arithmetic reasoning, attributing to it a somewhat different status than reasoning that is mediated by simple operators. The more general phrase "arithmetic reasoning" recognizes the fact that simple operators constitute only part of our ability to reason arithmetically, that is, to mentally manipulate representations of numerosity.

In certain circumstances, evidence concerning an individual's ability to reason about number cannot be interpreted without evidence about how he abstracts a numerical representation. Consider, for example, the theater usher who tells the manager that 200 people are crowded around the theater door. After he gets them all lined up in an orderly fashion, however, he remarks to the ticket seller that there are no more than 150. Are we to conclude that the usher lacks a number-invariance principle? Does he think that arranging a crowd of people into a line reduces their numerosity? An unlikely conclusion. If we were to ask the usher about his discrepant judgments he would no doubt point out that the crowd looked bigger when gathered around the door but that when they lined up it became obvious that they were not so numerous as he had thought. The usher's number-invariance principles were not manifest in his behavior, because the procedure by which he obtained a representation of the crowd's numerosity was not one in whose accuracy he had any confidence. His statements reported *independent* estimates of the crowd's numerosity. He was unperturbed by the discrepancy between the two estimates, not because he lacked number-invariance rules but rather because he had little faith in his initial representation of the numerosity.

A Martian investigating numerical reasoning in adult humans might erroneously conclude that his subjects lacked fundamental principles such as number invariance under displacement if he did not take care to establish that the humans were in a situation in which they could accurately represent the numerosities. The answers humans give to numerical questions may be only as good as the numbers (representations of numerosity) that serve as inputs to the reasoning principles. The maxim "Garbage in, garbage out" may apply as well to human numerical reasoning as to computerized data processing. We cannot properly begin an inquiry into a child's numerical reasoning principles (or those of adults in another culture) until we have established under what circumstances, if any, the child can accurately represent to himself the numerosity of a set.

Number Abstracting by Young Children

Considerable evidence exists that preschool children can identify the number of items in an array when the set size of the array is relatively small (Beckmann, 1924; Descoeudres, 1921; Gelman, 1972a; Gelman and Tucker, 1975; Lawson, Baron, and Siegel, 1974; Smither, Smiley, and Rees, 1974). Using a variety of tasks, such as discrimination learning, absolute judgment, and matching to sample, these investigators have found that children of preschool age can respond to the numerical value of a display and ignore such nonnumerical properties as length, density, item type, item color, and so forth. But, to repeat, this ability only holds when the set size of the display is small—somewhere between one and four or five items (Gelman, 1972a).

The most systematic studies of the effect of set size on the young child's ability to abstract numerosity were conducted by Beckmann and Descoeudres. Each of these investigators tested a large sample of children on a battery of tests, and both noted a drop in performance accuracy when sets of more than three items were used. The consistency of this result led Descoeudres to describe what she charmingly called the *un, deux, trois, beaucoup* phenomenon in preschool children. She used this phrase to indicate that the preschool child's ability to determine number appears to break down rather abruptly at some number between two and five, most typically after three in children younger than age 4. Descoeudres found that the typical 3½-year-old child could determine the numbers one, two, and three reliably in a variety of tasks; could represent the number four, but only imprecisely and unreliably, sometimes producing five or three objects instead of four; and was grossly inaccurate and unreliable with numbers greater than four. She introduced verbal protocols and other observations to demonstrate that the children regarded the larger numbers as all equal to "a lot" and therefore undifferentiated from one another. Still, as we shall see, it is not clear that young children fail to differentiate large sets from each other.

Beckmann credited children with having mastered a given number only if they had correctly judged that numerosity in a wide variety of different arrays, including some arrays composed of heterogeneous items. Thus, we can conclude from Beckmann's work that children can abstract numerosity correctly whether the array consists of homogeneous or heterogeneous items. This leaves open the question of whether children can more readily represent number when the numerosity to be represented consists of homogeneous items. The claim that this is so was advanced by Gast (1957) and has received some further support from the work of Siegel (1973). However, Gast's finding that preschoolers judge number more accurately for homogeneous arrays than for heterogeneous arrays may reflect his particular methodology. He tested the children on the homogeneous arrays first, a procedure that may have led them to assume that the task involved determining the numerosity of homogeneous groups of items. When subsequently confronted with the heterogeneous arrays, the children may have been uncertain of what was expected of them: Should they count only homogeneous groups of items or should they ignore differences between items and count everything together?

Siegel assessed the ability of young children to solve a discrimination-learning task on the basis of a numerical criterion, testing some children on heterogeneous arrays and others on homogeneous arrays. The dependent variable was trials to criterion. Children in the heterogeneous condition took longer to reach criterion; but they did reach it. As we pointed out earlier, younger children appear to be more inclined than older children to attend to irrelevant cues or dimensions when solving simple discrimination-learning tasks. This tendency slows down their rate of learning. Perhaps the reason young children perform less well with heterogeneous arrays than with homogeneous ones is that heterogeneous arrays provide more irrelevant cues to distract them. We will say more about this possibility in later chapters. For now, we summarize by saying that young children are able to abstract number from heterogeneous sets—although they do not always do so. The challenge is to determine what factors interfere with this ability.

In summary, earlier research has established that children as young as 2 years can accurately judge numerosity provided that the numerosity is not larger than two or three. Slightly older preschoolers can handle numerosities of three and sometimes four and five. Beyond five, the accuracy of numerical judgments falls off markedly. This earlier work leaves two questions unresolved. First, does the young child represent numerosities greater than three, four, or five as undifferentiated *beaucoups?* Second, does item heterogeneity have any effect on number judgments as such (rather than on the clarity of the child's understanding of the experimental task)?

Gelman and Tucker (1975) addressed themselves to these two questions. They showed cards to children between the ages of 3 and 5 and asked the children to indicate how many "things" they saw on each card. The cards had either homogeneous rows of stars or circles or heterogeneous rows consisting of red and blue stars and circles. Each child participated in either the homogeneous or the heterogeneous condition. Within these conditions each child saw cards depicting set sizes of 2, 3, 4, 5, 7, 11, and 19. Each set size was presented three times, once for only one second, once for five seconds, and once for one minute. Thus the experiment varied age, number to be estimated, homogeneity or heterogeneity of items in the array, and duration of exposure. Exposure time and number of items were within-subject variables; item type (homogeneous or heterogeneous) was a between-subject variable.

Table 5.1 summarizes the overall results of this experiment. Note that it includes results for set sizes larger than five, results that were not reported in the article by Gelman and Tucker. We will discuss the results of the exposure variable in Chapter 6 when we take up the question of what processes the young child uses to abstract numerosity. For now, we draw attention to the fact that children perform best overall when the exposure time is longest—a result that confirms previous findings in the literature. (See Gelman, 1972a, for a review of this variable.) Accordingly, when we discuss the ability of young children to abstract numerosity, we will focus on their performance on the one-minute exposure time. This focus is consistent with our emphasis on discovering what young children can do under optimal conditions.

Gelman and Tucker found the homogeneity-heterogeneity variable to have no effect under any conditions. This result agrees with the Beckmann and Descoeudres findings in that in those studies the children who passed a task did so across a range of stimulus conditions.

TABLE 5.1. Number of subjects who gave correct absolute judgments of numerosity.

		Set size						
Age group and exposure time	*N*	*2*	*3*	*4*	*5*	*7*	*11*	*19*
3-year-olds	48							
1 second		33	28	9	8	1	0	0
5 seconds		41	38	21	16	10	1	0
1 minute		41	40	28	27	20	16	5
4-year-olds	48							
1 second		44	37	23	17	4	2	1
5 seconds		44	41	29	21	11	3	1
1 minute		45	42	37	32	19	19	7
5-year-olds	48							
1 second		47	43	33	23	9	6	1
5 seconds		44	44	37	26	19	8	2
1 minute		47	46	42	38	27	19	8

For details of the procedure used to obtain these data, see Gelman and Tucker (1975). The data for set sizes 2–5 were reported in that paper; those for set sizes 7, 11, and 19 were reported for the first time in Gelman (1977).

Gast's data, on the other hand, suggest that heterogeneity has an effect and that the effect decreases as children get older. He reported that preschoolers accurately assigned cardinal numbers to small sets (sizes one to five) of homogeneous items but failed to do so when confronted with sets of items that varied in color, size, or shape. The difference between Gast's results and those of Gelman and Tucker may spring from differences in procedure. Gelman and Tucker tested each child on either homogeneous or heterogeneous stimuli. Gast tested each child on both types of materials, with the homogeneous materials presented first. As we mentioned earlier, Gast's children may have learned to expect homogeneous arrays, and this expectation may have masked their ability to judge the numerosity of heterogeneous arrays. A second study reported by Gelman and Tucker (1975) provides some evidence that expectations can interfere with children's ability to cope with heterogeneous arrays. They found that 3- and 4-year-old children behaved as if the numerosity of a homogeneous array changed when the experimenter removed one item and replaced it with an item of a different type. However, when children were presented first with a heterogeneous array and then with a homogeneous array of the same number, they did not seem to think the numerosity had altered. The verbal protocols of children in the two different conditions give a clue to the reason for this differential effect. Homogeneous items are described in terms of a specific referent ("there's three mice"); heterogeneous items are often described as *things* ("there's three things"). This suggests that heterogeneity affects the young child's ability to abstract a numerical representation by making ambiguous which items are to be counted together. The more the child is encouraged to define a collection of objects as *things,* the more obvious will be his ability to represent the number of items in a heterogeneous array.

What is the effect of set size in the Gelman and Tucker experiment? At this time we wish to examine only the data for the one-minute exposure time, since these data reflect the children's optimal level of performance. Note that for set sizes of 2 and 3 even 3-year-olds gave accurate answers. When set size rose higher than 3 the number of children who gave accurate answers declined for all ages—though less abruptly for older children. So far, the effect of set size is like that found by Beckmann and Descoeudres: The accuracy of numerical estimation in preschoolers falls off as numerosity becomes larger than 3 to 5. However, inspection of Table 5.1 suggests that numerosities greater than 5 do not constitute undifferentiated *beaucoups,* even for

3-year-olds. Preschoolers as young as 3, given sufficient time, clearly do better than chance when estimating numerosities as large as 19. Indeed, for numerosities as large as 11, between 30 percent and 60 percent of preschoolers' judgments are accurate, with the older preschoolers doing systematically better at all numerosities.

The reader may wonder how we estimate the chance level of performance under these circumstances. A really serious attempt to do "null hypothesis rejection" statistics on the percentages in Table 5.1 would presumably involve comparing children's conditional probability of giving *eleven* as the answer when confronted with an 11-item array to their unconditional probability of giving *eleven* as the answer. We have not attempted to do this. Instead we have regarded the percentage correct with one second of exposure time as a conservative empirical indicator of the chance level of accuracy. Our conclusion that even 3-year-olds are above the chance level of accuracy in estimating numerosities of 7, 11, and 19 derives from a comparison of the one-minute data with the one-second data.

Another way to determine whether or not preschoolers differentiate the larger sets is to consider the range of the answers given for each set size. What is the tendency for young children to represent larger set sizes in terms of number words that come later in the list of counting words? Do children, for example, tend to use the number words one, two, three, and four for representing small set sizes (say 1 to 5 or 6) and ten, eleven, and twelve for larger set sizes (say 7 to 19)? If so, we can conclude that young children have (*a*) some notion of the fact that the serial list of number words represents larger and larger sets, (*b*) some ability to differentiate set sizes larger than 5, and (*c*) some ability to map the ordinal properties of set size to serial position in the series of number words. It is important to note that the children in the Gelman and Tucker experiment were presented with the different set sizes in random order. Subjects were not shown a set size of 2, then 3, then 4, then 5, and so on. Thus any correlation between set size and number word used cannot be due to children simply starting with small number words and proceeding to larger and larger ones.

What number words do young children most often use to represent set sizes of 2, 3, 4, 5, 7, 11, and 19? How variable are these tendencies? To answer the first of these questions we determined the median of the answers given by each age group for each set size. To answer the second question we considered the range of number words used to represent a given numerosity. The complete range is not a meaningful index, however, since an occasional child said the set of size 19 was "a million" or "a zillion." In order to exclude these unmanageable guesses, our analysis focuses on the range within which 80 percent of the answers fall. The upper and lower bounds reported in Figure 5.1 actually exclude the highest 10 percent and the lowest 10 percent of the children's answers.

As Figure 5.1 shows, 3-year-olds tend to use the words two, three, four, five, six, ten, and eleven to represent the set sizes of 2, 3, 4, 5, 7, 11, and 19 respectively. In other words, larger numerosities tend to be represented by number words that come later in the sequence of counting words. By age 4, subjects tend to use the number word that accurately represents the number of items in an array. The median response of 4-year-olds and 5-year-olds to a 19-item array is nineteen. Many children whose answers are not accurate do nonetheless systematically increase their estimates in response to increasing set sizes, as can be seen from the systematic increase in the upper bound of the answers.

Note that in Figure 5.1 the lower bound of the estimates from the 3-year-olds remains at three across set sizes ranging from 3 to 19. This suggests that some of the subjects in this age group may be showing the *un, deux, trois, beaucoup* phenomenon mentioned by Descoeudres. That is, they may be treating set sizes of 3 and larger as undifferentiated *beaucoups,* to be labeled with the largest number word they know. To determine the extent to which 3-year-olds (and older children for that matter) confirmed Descoeudres's observation, we coded each child's answers. This process yielded six discernible patterns of responses.

Five of the patterns indicated a differentiation of the larger set sizes, accompanied by a more or less successful effort to use large number words to represent large numerosities. We termed the paradigmatic and most prevalent instance of such a pattern the perfect ascending sequence (*A*). In this pattern the rank order of the children's answers was perfectly correlated with the rank order of the numerosities. For example, a child would fit this pattern if he used the words two, three, four, five, seven, eleven, and twenty-three to represent the set sizes 2, 3, 4, 5, 7, 11, and 19 respectively.

Four of the other patterns resembled the perfect ascending sequence but included limited degrees of sloppiness in the correlation between the rank order of number words and the rank order of numerosities. We coded a pattern as ascending with one tie (AT_1) if the same number word was used to represent two adjacent numerosities, for example, if two, three, four, four, five, six and eighteen were used

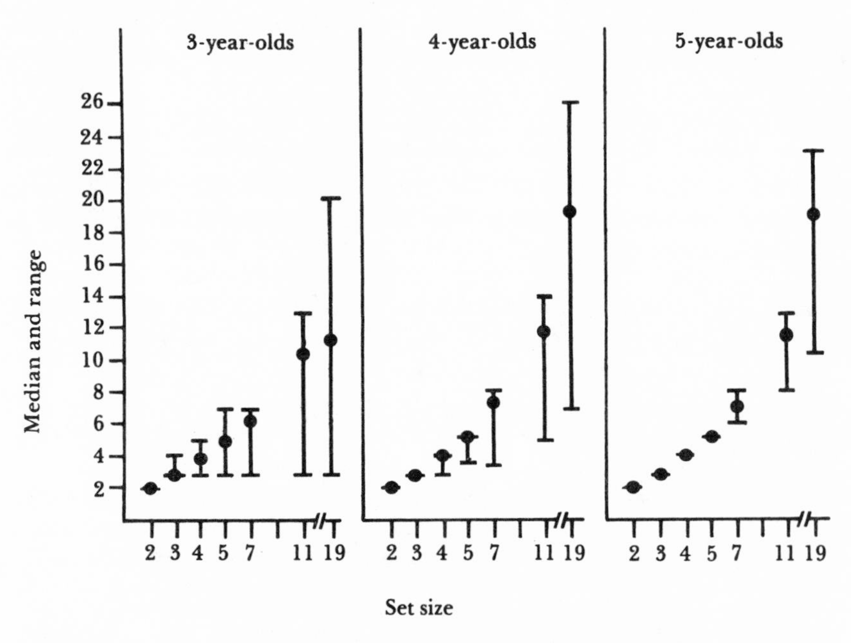

Figure 5.1. Median and range of preschoolers' estimates of numerosity, as a function of set size. The dots are the median estimates. The indicated ranges include all but the highest 10 percent and the lowest 10 percent of the estimates. Estimated numerosity clearly increases with actual numerosity, even when actual numerosity exceeds seven.

to represent set sizes 2, 3, 4, 5, 7, 11, and 19 respectively. A pattern was coded ascending with one reversal (AR_1) if two number words were reversed with respect to the rank ordering of the numerosities they were meant to represent, for example, if the set sizes listed above were represented by the words two, three, four, six, four, eleven, and nineteen. The other two patterns had either one reversal and one tie (AR_1T_1) or two reversals and one tie (AR_2T_1). All these basically ascending patterns, no matter how much they departed from perfect rank-order correlation, used at least two distinct number words beyond five to represent the numerosities 7, 11, and 19.

The sixth discernible pattern conformed more or less to what one would expect of children who regarded the larger numerosities as undifferentiated *beaucoups*. In this pattern, a single number word was used to designate nearly all of the numerosities. All of the following sequences are examples of this sixth pattern: (*a*) two, four, four, four, four, four, four; (*b*) five, five, five, five, three, three, three; (*c*) two, three, three, five, five, five, five; (*d*) two, three, three, three, three, three, twelve. Note that examples (*b*) and (*d*) do not conform very closely to the *un, deux, trois, beaucoup* model. They might best be regarded as instances of low-number-word "noise," since some are not systematic even in the Descoeudres sense. Nevertheless we include them here because they might be thought to suggest that the child regards the larger numerosities as undifferentiated. It should be noted that this method of assessing the child's representation of the larger numerosities is biased toward the conclusion that he does not differentiate them. The child may use the number words three, four, and five repeatedly simply because these are the only number words he knows rather than because he regards the 7-item array as being as numerous as the 19-item array.

This bias in the coding method makes it all the more remarkable that 65 percent of the 3-year-olds give evidence of differentiating between 5, 7, 11, and 19. As Table 5.2 shows, 31 of the 48 3-year-olds

TABLE 5.2. Number of subjects who used each discernible pattern of number words.

Pattern		Age group		
		3-year-olds (N = 48)	4-year-olds (N = 48)	5-year-olds (N = 48)
Perfect ascending	(A)[a]	17	20	30
Ascending with one tie	(AT_1)[b]	8	11	10
Ascending with one reversal	(AR_1)	3	5	6
Ascending with one reversal and one tie	(AR_1T_1)	2	1	1
Ascending with two reversals and one tie	(AR_2T_1)	1	0	0
Descouedres-like patterns	(D)[c]	12	3	0
Other (uncodable)		5	8	1

a. A = perfect correlation between rank order of number words and rank order of set size.

b. AT_1, AR_1, AR_1T_1, AR_2T_1 = various degrees of deviation from perfect rank ordering. To be credited with any of these patterns a child had to use at least two number words beyond five to represent set sizes 7, 11, and 19.

c. Failure to use number words beyond five to represent large set sizes.

used one of the ascending patterns; only 12 from this age group behaved in some way like Descoeudres's subjects. And by age 4 the vast majority (77 percent) are clearly able to distinguish set sizes between 5 and 19.

This evidence that preschoolers are able to differentiate set sizes of 5, 11, and 19 also illustrates another facet of preschooler's capacity to represent number. The use of ascending patterns reveals an implicit recognition of the fact that the rank order of the number words corresponds to the rank order of the set sizes—in the higher range as well as in the lower range. Thus the children reveal some understanding of the property of ordinal representation that is embodied in the sequence of number words, despite the fact that they may not be able to assign *the* number word that accurately represents a given set size.

In summary, Gelman and Tucker's data on the preschooler's ability to make absolute judgments of set sizes from 2 to 19 yield a general pattern of results like those reported by Beckmann and Descoeudres. Accuracy of judgments is or can be independent of item type within a set. As set sizes get larger, especially beyond a size of 5, the children's answers tend to be less accurate. The pattern of answers across the range of set sizes leads us to modify a standard interpretation of the young child's failure to give accurate answers when asked to judge large set sizes. The answers may be inaccurate, but they are orderly: Larger set sizes are represented by number words different from those at the very beginning of the sequence of number words. Further, the young child reveals some knowledge of the ordinal properties of number and their verbal representations. To our knowledge, this is the first demonstration that preschoolers are sensitive to the ordinal characteristics of larger numerosities *and* the ordinal characteristics of the number-word sequence *and* the conventional relation between the two. The young child's sensitivity to the order relation in itself has been demonstrated many times, particularly as it applies to continuous quantities such as length and weight (see, for example, Brainerd, 1973; Bryant, 1974; Bryant and Trabasso, 1971; Riley, 1976; Siegel, 1973; and Trabasso, 1975). Siegel (1973) has demonstrated an ability of children to learn to respond to the relative order of dot arrays representing the numbers two through nine. However, she provides no evidence on the preschooler's ability to differentiate larger set sizes or to use number words.

Our analysis of the children's pattern of responses to increasing set sizes not only shows a need to modify the general view that young children only know about small numbers but also prompts us to ask what processes mediate the young child's ability to represent numerosity. How does a young child learn to represent larger set sizes by number words that come later in a serial list? An obvious possibility is that such learning comes from practice at counting. But this assumes that preschoolers count in order to represent number—an assumption that is by no means prevalent in the literature. To answer the question we must examine both the hypotheses and the evidence about the nature of the preschooler's number-abstraction processes.

Chapter Six

How Do Young Children Obtain Their Representations of Numerosity?

The fact that young children can accurately represent number only when the numbers involved are small has led to the suggestion that the young child's concept of number is "intuitive" (Piaget, 1952) or "perceptual" (Gast, 1957; Pufall, Shaw, and Syrdal-Lasky, 1973). As we understand it, this position has two facets. First, it includes the view that young children obtain their representation of the number of items in an array by means of a direct perceptual-apprehension mechanism, sometimes referred to as *subitizing* (see, for example, Jensen, Reese, and Reese, 1950; Klahr and Wallace, 1973; Neisser, 1966; Woodworth and Schlosberg, 1954, Ch. 4; Schaeffer, Eggleston, and Scott, 1974). Second, it includes the idea that young children have yet to develop an ability to reason about number: to understand, say, that some transformations do not change the numerosity of a set (Piaget, 1952). This chapter is devoted to a consideration of the first of these views. Chapter 10 takes up the second one.

Young children's limited ability to abstract numerosity has served as the point of departure for theories of how they do what little they do. Over the years a variety of investigators have drawn attention to the ability of birds and various other animals to behave as if they too recognize the differences among numerosities up to three or four. For many years it has been assumed that "primitive" tribes are likewise limited in their ability to represent numerosity (Dantzig, 1967; Menninger, 1969). The observation of similarly limited abilities has led to the postulation of a common *low-level* process that mediates this ability. Many theorists consider perceptual processes to be in some sense low-level processes. Birds can form perceptions but lack higher mental abilities; young children have extensive perceptual abilities but very limited reasoning abilities; and "primitives" are like children insofar as their cognitive development fails to proceed to later stages (see Cole and Scribner, 1974, for a review of such theories). The common denominator that is assumed to cut across phylogenetic, ontogenetic, and cultural differences is perception. A bird, a young child, and an adult "primitive" share common perceptual processes—at least those that are required for the abstraction of a number-based representation of small set sizes. No one has provided a detailed account of the kind of perceptual process that might be involved. The process is generally thought to be simple: Each numerosity is grasped, apprehended, taken in as a whole, seen as a pattern. The idea is that there exist pattern recognizers that detect oneness, twoness, threeness, and so forth. Wundt (1907) made a similar suggestion about how we might process short temporal sequences. The rhythm of two taps is different from the rhythm of three taps. Whether the patterns being detected are in temporal rhythms or spatial arrays, small numbers are assumed to be processed as gestalten—distinct forms. *Twoness* and *threeness* are considered percepts like *cowness* and *treeness.*

We do not accept this argument. It takes the form of: Birds do it; young children and "primitives" do it. Since birds do it, it must be a low-level perceptual process. For one thing, we see no reason to assume that one common mechanism accounts for observed similarities in the behavior of birds, children, and "primitive" adults. Furthermore, the argument is circular. Animals and young and "primitive" humans must use low-level mental processes. Animals and young and "primitive" humans make numerical judgments. Therefore, the numerical judgments are based on low-level processes. Early in this book (Chapter 1), we outlined our objections to the broad assumption that children of preschool age lack all the sorts of cognitive competence that are assumed to require higher mental processes. Cole and Scribner (1974) discuss the pitfalls of applying such assumptions to observed cross-cultural differences in performance on cognitive tasks. And Zaslavsky (1973) provides ample evidence against ascribing a "one, two, many" type of arithmetical thought to so-called primitive African tribes.

Finally, even if we grant that a limited capacity for numerical abstraction characterizes a stage of development, it is not obvious that the pattern recognition is simple or low level. Efforts to model pattern perception highlight the complexity of the processes that must be involved in our ability to recognize shapes and separate figure from ground (Raphael, 1976). Several theorists treat perception much like cognition (see, for example, Bruner, 1957; Neisser, 1966).

By now the reader may well wonder why the assumption that small numbers are apprehended or subitized is so widespread. The argument based on comparisons between young children, birds, and "primitives" is far from convincing. We suggest that the hypothesis that young children subitize small numbers derives its main justification from the conviction that adults do. What support exists for this conviction? Klahr (1973) summarized evidence from experiments on the ability of adults to judge set sizes (for example, Jensen, Reese, and Reese, 1950; Kaufman et al., 1949; Saltzman and Garner, 1948). Klahr drew attention to two features of these studies (see Figure 6.1). First, adults respond more quickly to small set sizes than to set sizes

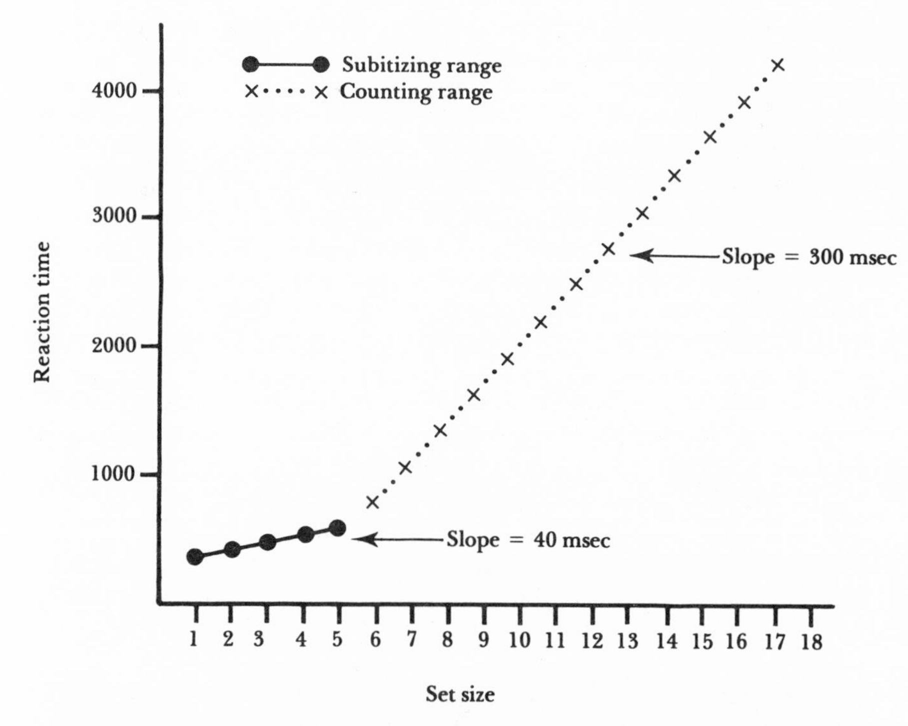

Figure 6.1. Time required for adults to accurately estimate numerosity, shown as a function of set size. For set sizes of five or fewer the reaction time increases by about 40 milliseconds for each additional item in the set. For set sizes greater than five, the increase per item is approximately 300 milliseconds. The acutal data are not as perfectly linear as this idealization suggests. (Based on Klahr, 1973.)

larger than five or six. Second, when asked to give accurate numerical judgments as rapidly as possible, adults seem to treat the small and the large set sizes differently. Reaction time (*RT*) functions for the two ranges are different. The slope of the *RT* function for small sets is decidedly shallow; each one-item increase in set size increases reaction time by approximately 40 milliseconds. In contrast, for set sizes larger than five the slope is approximately 300 milliseconds.

Klahr and Wallace (1973), as well as others (such as Neisser, 1966; Woodworth and Schlosberg, 1954), have inferred from this difference in slopes that small set sizes are processed by a subitizing mechanism and larger sets by a counting procedure. According to Klahr and Wallace, the slight increase in time needed to respond to each successive set size up to five reflects the time needed to retrieve the verbal label from the serial list of number words. They assume that the time taken to obtain a nonverbal numerical representation of the set sizes within this range is constant. Presumably, then, the representation of numbers within this range is accomplished by a direct apprehension. For the representation of larger numbers, Klahr and Wallace postulate the use of a counting mechanism. The subject notices the items in an array one at a time and pairs each item with the appropriate number word. When all the objects have been noticed, the last name assigned is used to represent the numerosity of the array. The one-by-one counting produces the steady slope of the reaction-time function for larger numerosities.

Klahr's explanation (1973) of the way adults make numerical judgments depends on two assumptions: that *RT* functions for small and large set sizes are notably different, and that the slope of the *RT* function for small set sizes is decidedly shallow. Questions have been raised about the veracity of these assumptions, especially the second (Allport, 1975; Weinstein, 1977). For purposes of the present discussion, however, we will assume that Klahr is correct about the way adults make numerical judgments. For accounts like Klahr's contribute to the developmental hypothesis under discussion: that young children are restricted to the use of a perceptual mechanism when they judge numerosity.

What does acceptance of Klahr's ideas lead us to think about the young child's ability to represent numbers? Recall that young children accurately abstract the numerosity of small sets but rapidly lose accuracy as set size becomes greater than four or five. The fact that young children's number-abstracting ability breaks down at about the point where adults appear to shift from subitizing to counting leads to the conclusion that young children subitize rather than count. We might further assume that young children have difficulty with larger sets because they cannot count. If so, we might expect to find that young children subitize before they count a given number. Further, we should find that young children seldom, if ever, count to represent small numbers. Finally, the counting of small numbers should appear in connection with the emergence of the ability to count large numbers.

Is it the case that children do not count small numbers? Do children begin counting small sets only when they begin counting larger sets? Do children subitize small numbers before they count them? Several lines of evidence bear on these questions—some more direct than others, some more compelling than others. Taken together, they suggest that the answer to all three questions is no.

To us, the most convincing evidence is Gelman's observations of young children in a variety of number experiments. She has found that counting is a salient behavior whenever the experiment permits. Indeed, it was the prevalence of spontaneous counting behavior that alerted her to the role counting might play in the way young children think about number. Excerpts of protocols that illustrate these observations are scattered throughout this book. The following protocol from a "magic" experiment (see Gelman, 1972b) is a particularly charming case in point.

S. M., age 3 years, 6 months, participated in an experiment that required her to label a linear display of five green toy mice the winner and a linear display of three green toy mice the loser. (Here and in all subsequent protocols, the child's words appear in italics.)

> Is that [the five-mouse display] the winner? *Yes.* Why? *Cuz it has one, one mouse; two, two mouses; three, three mouses; four, four mouses; five, five mouses.*
>
> [On a subsequent trial] Why does that [the five-mouse display] win? *Cuz it has five mouses. Count them! That* [the three-mouse display] *the loser.* Why? *It has three mouses. Let me count; one, two, three. Three! Mouses!*
>
> [Once again asked why the five-mouse display won, S. M. indicated some exasperation.] *Well, it has five mouses. Count.* [pause] *One.* [pause] *Say one!* One. *Now say two.* Two. *Say three.* Three. *Say four.* Four. *Say five.* Five. *See, five mouses.*

It is not only the magic experiments that provide a setting for observing the spontaneous counting behavior of young children. In several of the number-abstraction tasks employed by Gelman, children *ask* if they can count. If conditions permit (for example, in the long-exposure condition in the Gelman and Tucker experiment), the children count. Indeed, under conditions that tend to inhibit counting (such as short exposure times) young children often complain that they cannot count.

Such observations make it clear that young children can and do count small numbers. Note that S. M. even counts a three-item array. These data, however, do not tell us whether the children first subitize and then count. S. M. sometimes begins her answer by counting aloud, but at other times she first states the cardinal number representing the array and then counts.

To our knowledge, Beckmann (1924) was the first to suggest that the young child counts to represent a given small number *before* taking advantage of a subitizing or perceptual grouping process. This is the reverse of what should happen if Klahr and Wallace are correct. Beckmann cited the important fact that *the younger the child, the greater his tendency to count aloud* when answering questions about the number of items in an array. Gelman and Tucker (1975) report that young children are better able to estimate small sets under conditions that maximize their chances of using a counting procedure. When 3-year-olds are asked how many items there are in linear two-dimensional arrays of two, three, four, or five items, the longer the exposure time of the array, the greater their tendency to count aloud and the greater the accuracy of their answers. Increasing exposure time facilitates the judgment of small numerosities in a range of exposure times well beyond that required for a clear perception of the array.

Gelman and Tucker (1975) provide further evidence that number representations are first obtained by counting rather than by subitizing. They report an effect on accuracy scores caused by the interaction of set size, exposure time, and age. The 3-year-olds' ability to state the number of items in arrays of two, three, four, or five items increased with increasing exposure time—despite the fact that the shortest exposure time was one second, a time that presumably is long enough to allow subjects to form a well-defined perception of the array. Lengthening the exposure time beyond one second allowed the 3-year-olds to become more accurate at judging even sets of two and three. On these smallest sets, the 4- and 5-year-olds did as well with one second of exposure time as with longer exposure time. On sets of four and five, however, the performance of the 4- and 5-year-olds did improve with longer exposure time.

What to make of this interaction? We see two possible interpretations. First, it could mean that the form of Beckmann's hypothesis is correct: A number must first be counted, even if it is as small as two or three. Practice at counting allows the child to skip the counting process and "chunk" the array. Slightly larger arrays, which are not yet so readily chunked, still must be counted. Eventually, they too can be chunked. Note that this view of the process of abstracting number does not treat subitizing as a lower-order, primitive, simple mechanism. Instead it assigns subitizing an advanced organizing role. As Beckwith and Restle express it, "Perception of small numbers may be a skill developed by adults, a sort of shortcut to counting, rather than an elementary mental event" (1966, p. 439). The imposition of gestalt organization on the array makes it easier for the child, and the adult for that matter, to process yet larger arrays. This position admits that the very young child may use some perceptual process in abstracting a numerical representation. But rather than considering such perceptual activity as primitive, it treats it as an aid to counting.

A second interpretation of these data is that even 4- and 5-year-olds count set sizes of two and three but that they do so rapidly and subvocally. A study by Chi and Klahr (1975) provides some support for this view. In a tachistoscopic study of 5-year-olds, Chi and Klahr found that the mean reaction time for accurate judgments of three-item arrays was 195 milliseconds longer than the mean reaction time for accurate judgments of two-item arrays. Such a large difference in reaction time is at least consistent with the assumption that 5-year-olds count two- and three-item arrays, albeit subvocally. Furthermore, of estimation studies with older children, those that restrict exposure time consistently report lower accuracy than those that do not restrict exposure time (see Gelman, 1972a, for a review). Whatever distinguishes twoness from oneness and threeness from twoness takes longer to "apprehend" for higher numerosities. One straightforward interpretation of these data is that numerical representations of numerosity are always formed through a process of serial enumeration (counting), whether that process is vocal or subvocal, conscious or unconscious. Perhaps twoness and threeness are not like cowness and treeness and can never be simply apprehended.

We do not wish to choose between perceptual chunking and rapid subvocal counting. It is easy to imagine how both types of processes—perceptual and motor—could be used to chunk small sets of objects; we shall return to this in Chapter 12. For now, we point to the evidence showing that very young children count small sets and that they seem to prefer counting over noncounting processes. These data serve to answer the third question raised by the subitizing-first theory:

Do children count small sets only after they begin to count large sets? All the data point to the difficulty children have in counting large sets at an age when they clearly can count small sets. As we shall see in Chapter 8, 2- and 3-year-old children willingly count displays containing 2, 3, or 4 items but refuse to count displays containing 7, 11, or 19 items.

Granting the availability of a counting procedure to young children allows us to explain some further results from studies of the young child's ability to abstract number. Gelman and Tucker found that heterogeneity of items in a perceptual array had no appreciable effect on the accuracy of number estimates, provided that the experimental design did not lead the children to expect only homogeneous groups of items. If the young child estimated small numbers on the basis of a direct perceptual apprehension of numerosity, one might expect heterogeneity to serve as a gestalt-destroying factor and to impede or prevent accurate estimation. With counting, however, which simply involves the serial tagging of the perceptual entities in an array, heterogeneity need not have an adverse effect on accuracy. We do not mean to suggest that no stimulus conditions would impede the tagging process. Rather, we wish to redirect analysis of the source of interference away from an exclusive focus on whether the stimulus is heterogeneous or not. Schaeffer, Eggleston, and Scott (1974) found that 4-year-olds were better at counting arrays composed of heterogeneous items arranged in homogeneous subgroups, than at counting entirely homogeneous arrays. As Beckwith and Restle (1966) pointed out, in counting one has to partition the items already counted from those still to be counted. Introducing heterogeneous subgroups probably facilitated the tagging process for Schaeffer, Eggleston, and Scott's subjects by giving them an external prop for partitioning the set. In view of our conjecture about the development of chunking strategies, it would be interesting to know both when children develop an ability to use such external props and when they outgrow the need for them by becoming able to generate them internally.

Despite all the evidence of the prepotent tendency of young children to count small sets, one could still hold the view that very young children process number by a subitizing mechanism. The reason? None of the data considered covers subjects younger than 3 years of age. Perhaps even younger children prefer to subitize. Schaeffer, Eggleston, and Scott made such a suggestion, pointing to the fact that 2-year-olds seem unable to count accurately and yet appear to make some number-based responses.

Perhaps at some stage children do subitize numerosity without being able to count. As the data we present in the next few chapters will show, such a stage, if it exists, must occur before 2 years of age. The rudiments of counting are clearly present in most 2-year-olds. More importantly, it appears that the child's arithmetic reasoning is intimately related to the representations of numerosity that are obtained by counting. The domain of numerosities about which the child reasons arithmetically seems to expand as the child becomes able to count larger and larger numerosities. Although it remains possible that children younger than age 2 can recognize differences in numerosity without counting, representations of numerosity obtained by direct perception do not appear to play a significant role in arithmetic reasoning.

Chapter Seven

The Counting Model

What does it mean to say that a young child *counts?* Adult counting is fairly transparent. Its essential features are readily discernible. It involves the coordinated use of several components: noticing the items in an array one after another; pairing each noticed item with a number name; using a conventional list of number names in the conventional order; and recognizing that the last name used represents the numerosity of the array. Beckwith and Restle (1966), Churchill (1961), Klahr and Wallace (1973), and Schaeffer, Eggleston, and Scott (1974) all agree on this analysis. When we say that a young child counts, however, we do not necessarily mean that the child follows the adult pattern. We do not grant that a child can count if and only if he uses all the components of the counting process. We do not require that the tags the child uses to mark the items in an array be the conventional number words. Nor do we insist that the child who uses number words as tags use them in the conventional order.

The various count sequences we have recorded for young children lead us to believe that the young child's ability to count is governed by several principles and that adherence to some of these principles requires the coordination of several component processes. We build our explanation of the counting procedure and its development by introducing the principles one by one and examining the processes that underlie adherence to each principle. In presenting a principle-by-principle model of counting we explicitly recognize that children may possess some counting principles but not others; that individual principles may draw on component skills, some of which may not be perfected at a given age; and that some of the principles may operate more or less in isolation in the counting behavior of very young children. In the end, of course, successful counting, involves the coordinated application of all the principles. However, discussions of development that focus only on the appearance of completely accurate counting underestimate the young child's knowledge of the counting procedure. Partial competence or a limited ability to coordinate the counting principles is not the same thing as a complete lack of competence.

Some investigators (such as Schaeffer, Eggleston, and Scott, 1974) report that 2-year-olds do not count and lack some or all of the counting principles we find in children of this age. Many of the earlier analyses required correct performance with the traditional sequence of number words. That is, the child's ability to count was measured against the adult's way of doing so. We have already discussed our objections to the use of adult criteria as the basis for deciding whether or not to credit young children with a given capacity. Our objection to requiring the use of the conventional order derives from the realization that it is possible to count without using the standard count words as tags. We suspect that a failure to recognize this fact contributed to the widespread belief that the members of many African tribes cannot count. The absurdity of this belief has been amply demonstrated by Zaslavsky (1973), who not only shows that Africans do indeed count and have done so for centuries but also provides a wealth of clues as to why so many investigators have failed to discover this counting ability. We consider some of these clues here, because they reinforce our point that requiring the correct use of conventional number words can be a dangerous business.

First, many African societies (for example, the Kinga, Hehe, and Nyatura of East Africa) use finger gestures and hand configurations to represent different set sizes. The gesture system does not always agree with the locally spoken number words. Indeed there need not be a corresponding system of number words. A failure to recognize that gestures may be used as tags in enumeration would necessarily lead to an underestimate of the extent to which such a society can count.

Second, number-word sequences need not derive from a base-ten system. Yes, English uses a base-ten system, but count-tag sequences can be and frequently are derived from other bases. Consider the fact that computers count with a base of two. There is a Bushman language that combines the words for one and two to get the words for three and four. The comparable English count-word sequence for one through four would read "one, two, one-two, two-two." If we failed to realize that the Bushmen were employing a binary concatenation rule, we might conclude that they could only count up to two.

A breathtaking instance of this form of numerical ethnocentrism appears in Menninger (1969): "The number sequence begins with three: 'three, four, five . . . etc.' When a tribe of South Sea Islanders counts by twos, *urapun, okasa, okasa urapun, okasa okasa, okasa okasa urapun* [that is, 1, 2, $2^{1}1$, $2^{1}2$, $2^{1}2^{1}$ 1], we distinctly feel that they have not yet taken the step from two to three. And we realize with astonishment that these people can count beyond two without being able to count to three" (p. 17). And Menninger is a comparative linguist!

Third, in a large number of East African cultures the counting of cattle, people, and valuable possessions is taboo. This cultural fact could lead even the best-intentioned field worker to conclude that the members of such cultures cannot count. It is easy to imagine that the field worker would take the precaution of asking individuals to count familiar items. And what is more familiar than the cattle, people, and houses in the village? Yes, they are familiar but they are also the very items that must not be counted! Questions like "How many children do you have?" are typically evaded. Sometimes an intentionally wrong answer is given. It seems that an erroneous count is one way to circumvent the taboo. Zaslavsky (1973, p. 255) tells of an American teacher of Kikuyu children in Kenya who asked her class to count the number of children in class that day. There were actually 26 but the children said 25—one short of the actual number. Zaslavsky suggests that the children made the error on purpose. By failing to give the exact number they avoided appearing to count. Another possibility is that the children did not count themselves (compare Piaget, 1928). For either of these reasons, it seems unreasonable to conclude that the children could not count just because they gave the wrong final number word.

Parenthetically, a recognition of the taboo against counting cattle helps provide an interpretation of another common assumption about Africans, namely that they are endowed with inordinate perceptual capacities as compensation for their limited capacity for abstraction. Not being allowed to count cattle presents a formidable problem for people who measure their wealth by the size of their herds. Throughout Africa, a variety of devices have been used to circumvent the taboos. The common feature of these devices is the establishment of a one-to-one correspondence between a collection of nontaboo items, which can be counted, and a collection of taboo items, which cannot be counted. Counting the nontaboo items thus makes it possible to represent the number of the taboo items. But what to do in everyday life when it is necessary to know whether any cattle have gone astray or whether a neighbor's cattle have joined your herd? Learn to recognize each and every member of your own herd. According to Kenyatta (1953), this is precisely what the Kikuyu do. Indeed, young Kikuyu boys are put through a rigorous perceptual learning experience. To be sure the young boys have learned their lessons well, the elder men mix herds and hide animals from a given herd until the boys can tell which animals are missing or do not belong to a herd. Gibson (1969) describes a similar lesson in perceptual learning regarding the way people come to identify individual members of a goat herd. It seems to us that the Kikuyu's perceptual prowess is just another case of learning to identify features that distinguish members of a set that appear alike to the untutored eye.

It is not just Africa that provides examples of the problems one can run into by relying on the ability to use a count-word sequence as a critical criterion for determining whether one can or cannot count. Many cultures have prohibitions against counting people. Orthodox Jews require that 10 adult men be present before they conduct a prayer service. The collection of 10 is called a *minyan.* In order to determine whether a minyan is present, the men recite a 10-word sentence, which contains no count words. Each man recites one of the words. If the phrase is completed, a minyan is present. Two phrases that are commonly used are *Hoshe'a et ameha, ovoeh et nahalateha, ooraam venasaam ad ha'olum* and *Boruch ata adonoi, alohanu melech ha'olum, homotzi lechem min ha'aretz,* the latter phrase being the blessing said over bread. The modern Orthodox Jews, like the African herders, avoid counting by setting up a one-to-one correspondence between the taboo items and some countable numerosity. Imagine how difficult it would be for someone unfamiliar with the rules and rituals of Orthodox Judaism to assess the congregation's ability to count!

The point should be made by now: There is no reason to require a child to use conventional count words in conventional order. What is it then that must be assumed? The use of unique tags to mark or tick off the items in a collection is intrinsic to the counting process. Further, the tags must be used in a fixed order. Finally, the tags must have an arbitrary status; they cannot be the names or descriptions of the items in the collection being counted. The set of count words meets these criteria, but then so do other sets of tags. One obvious candidate is the alphabet, and it is noteworthy that many languages have used the alphabet as count words (Greek and Hebrew, for example). But the tags need not even be verbal. They may be any of a host of entities, including short-term memory bins. Recognition of this fact leads us to introduce some terminology in order to be able to refer separately to the general category of possible count tags and the subset of such tags which constitute the traditional count words. We call the former *numerons;* the latter *numerlogs.* Numerons are any distinct and arbitrary tags that a mind (human or nonhuman) uses in enumerating a set of objects. Numerlogs are the count words of a language.

Note that in making this distinction we hold open the possibility that animals, insofar as they can be shown to base their behavior upon numerosity as such, use a counting procedure. That is, they may tick off items in an array, one by one, with distinct mental tags employed in a fixed order, and use the final mental tag as a representation of numerosity. This is nonverbal counting. Our terminology also allows for the possibility that young children use nonconventional or idiosyncratic tag sequences when counting, as indeed they sometimes do.

Since we do not insist on the use of conventional count words, we must define precisely what is involved in the counting procedure. We believe that five principles govern and define counting. In what follows we present each principle separately, because we analyze our data on the counting ability of young children by considering the availability of each principle as well as the ability to use the principles in concert. The first three principles deal with rules of procedure, or *how to count;* the fourth with the definition of countables, or *what to count.* The final principle involves a composite of features of the other four principles.

Counting Principles

The One-One Principle

Every counting model we know of assumes the use of what we call the one-one principle. The use of this principle involves the ticking off of the items in an array with distinct ticks (tags, numerons, numerlogs) in such a way that one and only one tick is used for each item in the array.

To follow this principle, a child has to coordinate two component processes: *partitioning* and *tagging.* By partitioning we mean the step-by-step maintenance of two categories of items—those that are to be counted (the set *U* in Beckwith and Restle's 1966 notation) and those that have already been counted (the set *C* in Beckwith and Restle's notation). Items must be transferred (either mentally or physically) one at a time from the to-be-tagged category to the already tagged category. Coordinated with this process is a second process that involves summoning up, one at a time, distinct tags (numerons). This process of course requires that the system have available a set of distinct tags. In this culture, the distinct tags typically used by adults are the counting words (numerlogs). But, as we have already seen, the numerons need not be the numerlogs. For a young child who is just mastering the counting procedure, it is not obvious that the tags do or should correspond completely to the traditional count words. In any event, in the successful use of the one-one principle, the summoning up of distinct tags must proceed in lockstep with the first process. As an item is transferred from the to-be-counted category to the counted category, a distinct tag must be withdrawn from the set of mental tags. By *withdrawn* we mean that the tag must be set aside, not to be used again in that particular counting sequence.

In other words, the one-one principle in counting requires the rhythmic coordination of the partitioning and tagging processes. The two processes must start together, stop together, and stay in phase throughout their use. Realizing that these processes must be coordinated leads us to ask whether young children employ strategies that serve the coordination. One such strategy would be to point to each item as it is counted. Pointing can serve to mark the withdrawal of a tag—especially if it occurs in conjunction with the verbalization of a numerlog—and at the same time help the child partition items that have been counted from those that have yet to be counted. Children's tendency to point when they count (see Chapter 8) provides convenient data for assessing the extent to which they recognize that the one-one principle requires coordinating the tagging and partitioning processes.

Within the one-one principle there is room for three kinds of errors: (1) errors in the partitioning processes such as ticking off an item more than once or skipping an item; (2) errors in the process of withdrawing tags such as using the same tag twice; and (3) failure to coordinate the two processes completely. As an example of incomplete coordination, the child may not stop withdrawing tags at the same time as he finishes transferring items into the already counted category. He may stop withdrawing tags either before or after he finishes transferring items. Also, he may not keep the tag-withdrawal process in step with the partitioning process, so that although the processes start and stop together the number of tags he withdraws is different from the number of items he transfers.

We know of no work that considers whether young children possess all components of the one-one principle or the extent to which they successfully integrate these components. Potter and Levy (1968) specifically addressed the question of whether children as young as 3 have what we call the partitioning component. They asked their subjects to touch each object in an array once and only once, and they found that the children were able to do so. Orderly arrays (such as ones arranged left to right) were more conducive to one-one pointing than were randomly arranged arrays. This result is what we would expect, because a linear array has a clearly marked beginning and end. If the child can move from left to right (or vice versa) through a linear array, he has a device for keeping track of tagged and to-be-tagged items. Potter and Levy's results suggest that we will find most young children able to partition to some extent. They leave open, however, the questions of whether young children can generate distinct tags and whether they can coordinate the partitioning and tagging processes.

The Stable-Order Principle

Counting involves more than the ability to assign arbitrary tags to the items in an array. Even if a child uses numerlogs as tags we cannot conclude that he necessarily knows the counting procedure. He must demonstrate the use of at least one additional principle—the stable-order principle. The tags (numerons) he uses to correspond to items in an array must be arranged or chosen in a stable—that is, a repeatable—order.

This principle calls for the use of a stable list that is as long as the number of items in an array requires it to be. This principle should present the child with some notable practical problems. It is well known that the human mind, unlike the computer, has great difficulty in forming long, stably recallable lists of arbitrary names (words). By arbitrary, we mean having no generating rules underlying the sequence. As we will indicate in greater detail in Chapter 12, we believe that much of the development of the young child's numerical abilities involves the rote learning of the first 12 or 13 number words and the generative rules for producing the subsequent number words. In other words, a significant part of the development of numerical abilities centers around the need to solve the practical difficulties posed by the stable-order principle. It makes sense to expect that the extent to which young children adhere to this principle is related to set size.

The Cardinal Principle

The two preceding principles involve the selection of tags and the application of tags to the items in a set. The cardinal principle says that the final tag in the series has a special significance. This tag, unlike any of the preceding tags, represents a property of the set as a whole. The formal name for this property is the cardinal number of the set. Put more informally, the tag applied to the final item in the set represents the number of items in the set.

So besides being able to assign numerons and do so in a fixed order, the child must be able to pull out the last numeron assigned and indicate that it represents the numerosity of the array. Insofar as the selection of one particular numeron and the use of it to designate the numerosity of the array requires additional processing steps, it seems likely that the cardinal principle has a developmental relationship to the one-one principle and the stable-order principle. The cardinal principle, which presupposes the other two, should develop later.

The Abstraction Principle

The three principles discussed so far describe the working of the counting process. They are *how-to-count* principles. The abstraction principle states that the preceding principles can be applied to *any* array or collection of entities. Note that we make no distinction between physical and nonphysical entities. Adults, at least, are able to count the number of minds in a room. Adults can also be induced to count ludicrously heterogeneous sets, such as the set consisting of all the minds and all the chairs in the room.

It is an open question whether young children realize that counting can be applied to minds, or pure products of the imagination, or even heterogeneous sets of objects. Ginsburg follows a long-established tradition in maintaining that early counting—and the concept of number itself—are "tied to particular concrete contexts, geometric arrangements, activities, people, etc.; it is a long time before the young child treats number as abstract" (1975, p. 60).

Gast (1957) devoted a lengthy monograph to experimentally documenting the supposed lack of abstractness in the young child's conception of number, concluding that for 3- and 4-year-olds,

> the strong dependency on composition and arrangement was equally evident in their simultaneous apprehension [subitizing] of small directly apprehendable numerosities and in their serial counting of greater numerosities. Enumeration is possible for 3- and 4-year-olds only when the things to be counted are identical to one another and arranged either in a group next to each other or in a homogeneous row. Any variation in material or arrangement has the effect of making

> the children enumerate in accord with the immediately apprehended obvious relationships. That is, elements that vary in material composition or qualities (such as color) are not included in the enumeration; irregular arrangement of the items and natural relationships between items lead to breaks in the smooth course of enumeration. (Gast, 1957, p. 66, Gallistel's translation.)

Gast concludes that only children 7 years of age or older have a fully abstract conception of what can be counted. As we will see, however, there is reason to believe that Gast underestimates the extent to which young children recognize the inherent generality of the counting procedure.

Note that for the purposes of counting, perceptual entities may be categorized in the most general way, as *things.* Even *things* may be too limiting a specification of what children and adults regard as countable. We have observed children counting the *spaces* between items in an array. From the standpoint of the adult, this extremely general category of what is countable may seem highly abstract. The view that the ability to classify physical events as things is indeed very abstract is implicit in theories of cognitive development that postulate an elaborate hierarchy of subcategories as the basis for categorization skills. Such a view is central to Gast's thesis that children slowly develop a more and more abstract conception of what constitutes a countable numerosity. A similar argument is found in Klahr and Wallace (1973).

Actually, one need not assume that a complex hierarchization scheme mediates the ability to classify entities as things. It is possible to view the ability to classify the world into things and nonthings as a derivative of the ability to separate figures from grounds. In this case, the categorization of things as opposed to nonthings may well be among the earliest (most primitive?) mental classifications. A differentiated and ordered hierarchy of subcategories of things may well be a later development. Thus if young children turn out to be able to classify heterogeneous items together for purposes of counting, we have an alternative to the conclusion that they have developed a complex generalization of the concept of thing.

As we have indicated, the abstraction principle is not a how-to-count principle. It concerns the range of applicability of the how-to-count principles. In other words, it is permissive but not restrictive as to what items may be counted. As adults, we recognize that *any* events or entities can be classified together for purposes of counting. These events can be arbitrary inventions of the mind, physical entities, sets themselves (as when one combines pennies into groups of five and then counts the groups), and so forth. The developmental question is what children of different ages consider to be the permissible range of things that can be included in an enumeration.

The Order-Irrelevance Principle

Consideration of the ways in which the principles outlined so far might interact leads us to postulate a final principle of counting, the order-irrelevance principle. This principle says that the order of enumeration is irrelevant; that the order in which the items are tagged, and hence which item receives which tag, is irrelevant. In other words, "It doesn't matter how you count."

Adults know that each of the count words can be assigned to any of the items in the array—so long as no count word is used more than once in a given count. That is, adults know that the order in which items are partitioned and tagged does not matter. Given a linear array of a rabbit, a truck, a dog, and a cat arranged left to right, adults understand that it is perfectly all right to call the rabbit *one* on one trial and the cat *one* on another trial. Ginsburg (1975) cites a number of anecdotes to demonstrate that young children lack this understanding.

The child who does appreciate the irrelevance of the order of enumeration can be said to know, consciously or unconsciously, the following facts: (1) that a counted item is a *thing* rather than a *one* or a *two* (the abstraction principle); (2) that the verbal tags are arbitrarily and temporarily assigned to objects and do not adhere to those objects once the count is over; and most importantly, (3) that the same cardinal number results regardless of the order of enumeration. In general, the order-irrelevance principle concerns the fact that much about counting is arbitrary. This principle is one of knowledge about the consequences or lack of consequences of applying the first four principles. At times it is difficult to draw a sharp distinction between principles of number abstraction and principles of number reasoning. This last principle deals not just with our ability to count but also with our understanding of some properties of numbers. The reasoning principles likewise deal with our understanding of the properties of numbers. Accordingly, we shall reintroduce the order-irrelevance principle in discussing the formal structure of the child's arithmetic reasoning (Chapter 11).

Chapter Eight

The Development of the How-To-Count Principles

We have called the one-one, stable-order, and cardinal principles the *how-to-count* principles. Our data on the development of these principles comes from two types of studies conducted by Gelman: "magic" experiments and videotaped counting experiments. Both types of studies are rich sources of observations about just how children count.

Gelman's magic studies, although not designed to measure counting ability, proved to be unexpectedly conducive to counting and to talking about number. Typically, Gelman's subjects counted—without being asked. We have no explanation of why they did so. The fact that they did provides us with what might be considered an optimal data base. Children who count spontaneously can hardly be said to be uninterested in counting. Presumably, then, the magic experiments show the children's counting behavior at its best. In view of our belief that students of cognitive development must discover what children *can* do, it seemed reasonable to begin constructing a counting model with such data. We recognized, however, that the children might perform differently if we asked them to count. This realization led us to run the videotaped counting experiments, which we devised specifically to test our model.

Evidence from Magic Experiments

Issues of Methodology

The magic experiment. The magic paradigm was designed initially as a test of the preschooler's ability to reason about number. In particular, the question was whether young children differentiate between two categories of transformations that can be performed on a set. Gelman sought to determine if young children treated addition and subtraction as number-relevant transformations and substitution and spatial displacement of items as number-irrelevant transformations.

She employed a two-phase procedure: The first phase established an expectancy for number; the second recorded the child's reaction to surreptitiously performed ("magic") transformations of the expected arrays. In the first phase children were shown two plates containing different numbers of small plastic toys. For example, one plate might hold two mice and the other three mice, or one might hold three mice and the other five mice. The experimenter designated one of these "the winner" by simply pointing to it—making no reference to number. The other plate was designated "the loser." Phase I then became an identification game. The plates were covered with cans and shuffled. If the child appeared to be keeping track of the covered winner, as one is able to in a shell game, the shuffling continued until the child appeared to have lost track. When the shuffling stopped, the child was asked to guess where the winner was and then uncover that plate to see if he was right. If he had guessed incorrectly and recognized that he had, that is, if he said that the uncovered plate was the loser, he was immediately allowed to uncover the other plate. When the other plate was uncovered, he was again asked if it was the winner. The experimenter provided immediate feedback about the child's identifications. If a child erroneously identified a loser as the winner, for example, the experimenter said, "No, that's the loser," and covered and shuffled the plates again. Whenever the child correctly identified the winner plate, the experimenter gave him verbal confirmation and a prize before recovering and reshuffling the plates. Note that the feedback from experimenter to child was based on the child's correct or incorrect identification of a plate after he had uncovered it, not on his guesses. Each uncovering of a plate was counted as a trial. Thus, when a child uncovered the loser plate, correctly identified it as such, and then uncovered the winner plate, the sequence counted as two trials. The purpose of running what was basically an identification experiment in the ostensible form of a shell game was twofold: First, the game-like nature of the task was extremely effective in engaging the young child's interest. Second, it built up a strong expectation about which display would be the winner and which the loser. This first phase of the experiment continued for at least 11 trials. On 3 of these trials, the child was asked to justify his identification. This served to determine what the child considered to be the definitive properties of the sets.

The second phase of the experiment began when the experimenter surreptitiously altered the winner set. In some experiments, the experimenter altered the spatial arrangement of the set, making it longer and less dense or shorter and more dense. In other experiments, the experimenter altered the color or identity of one of the elements. And in still other experiments, the size of the set was changed by adding or subtracting one or more elements.

From the child's standpoint, Phase II was just a continuation of Phase I until he discovered that neither plate contained a set that was identical to what had been the winner set. Again the child was asked to identify each plate as winner or loser. When the child had uncovered the altered winner plate, the experimenter asked a series of questions: Had anything happened? If so, what? How many objects were now on the plates? How many objects used to be on the plates? Could the game go on? Did the game need fixing? If so, how could it be fixed? If the child said that the game had to be fixed and that he needed certain items to fix it, he was given a handful of items that included the ones he had asked for plus several others.

Everything the child said was tape-recorded for later transcription. The experimenter also rated the degree of the child's surprise in Phase II on a three-point rating scale and made notes about any striking aspect of the child's behavior, such as search behavior.

The experiment was very effective at bringing out spontaneous counting and talk about number. The children talked about and used number not only at the end of Phase II, when they were asked to do so, but throughout the experiment. They frequently counted the items on the plates or stated their number, even when no questions had been asked; and they frequently counted the prizes they had won. The counting behavior observed at the beginning of the experiment was all the more remarkable in that the experimenter took pains to avoid mentioning number during the first phase of the experiment. The counting that occurred throughout the magic experiment —both before and after children were asked questions about number —forms the data base for our initial inferences about the nature of the child's counting process. These sequences were already present in the transcripts of protocols from the magic experiments. Most of the transcripts were made before we became interested in the analysis of counting, and all of them were made by transcribers who knew nothing of our interest in counting behavior. Therefore the transcribers are unlikely to have biased the transcripts.

The choice of protocols. In choosing the set of transcripts to use for our counting analyses we were guided by two considerations. First, we wanted to consider counting sequences generated by 2-, 3-, and 4-year-olds under the same experimental conditions. Children from each of these age groups had participated in magic experiments that displayed sets of two and three items in Phase I and then confronted children with the effect of either removing one item from the three-item array or displacing (so as to shorten or lengthen) the three-item array. Second, we wanted to consider the effect of set size. In a recent magic experiment, Gelman (1977) showed three- and five-item arrays to 3- and 4-year-olds in Phase I. Phase II displayed the effects of removing two items from the five-item array or displacing the items of the five-item array. Thus, subjects saw as many as eight items (three plus five) in this experiment—the largest number of items so far employed in Gelman's magic experiments.

The above considerations led us to focus on two sets of magic experiments: (1) those that employed two- and three-item arrays and compared the effects of subtraction and displacement; and (2) those that employed three- and five-item arrays and compared the effects of subtraction and displacement. Henceforth, we refer to these subtraction experiments as the [(2 vs. 3) − 1] and [(3 vs. 5) − 2] experiments. In both subtraction and displacement conditions, the items in the arrays were identical one to the other (for example, green toy mice) and arranged in linear displays.

The definition of a count sequence. For purposes of data analysis we identified a count sequence primarily on the basis of a child having used tag items from two well-known lists—the number words and the letters of the alphabet. A child did not have to use these tags in the conventional order, but he did have to use them.

The reader might be surprised to see that in this case we are relying on the use of conventional verbal tags. In Chapter 7 we pointed out that the ability to count should not be considered to be the same thing as the ability to use a culture's conventional count words in the conventional order. So why are we now restricting our analysis to count words and letters of the alphabet? The answer comes from the nature of our data base. If we had asked the children to count, we might consider for our analyses any behavior that followed the request, though we would be well advised to exclude certain obviously irrelevant behaviors (such as requests to go to the bathroom). But we did not ask children to count. This leaves us with the problem of knowing how to identify the beginning of a count sequence. Our guess that children were counting came from the observation that they, on their own, used strings of number words or letters. Thus, for these analyses, we decided to use the child's verbal behavior to index a count sequence. Note that despite our decision to rely on verbal behaviors and therefore, from our point of view, to be conservative in our definition of a count sequence, we do follow one of our caveats, namely that the child need not use tags in the conventional order. Thus, we analyze sequences like two, six; ten, two, six; and one, eight, five for what they might show about the child's ability to use counting principles.

Results

Evidence for the one-one principle. A particular count sequence was scored as evidence for the use of the one-one principle if it had the same number of different verbal tags as the number of items in a given array. Thus, if a child said two different number words or letters of the alphabet when confronted with a two-item array, he was scored as having demonstrated the use of the one-one principle on that trial. Likewise, if a child said three different number words when viewing a three-item array, the count sequence was so scored. Thus the child who said "two, six" or "*A, B*" when confronted with a two-item array was scored as correct (that is, correctly using the one-one principle), just as was the child who said "one, two." Further, a child who said "one, four" on one trial and "four, one" on a later trial was scored as correct for both trials. In other words, consistency with respect to order over trials was not a criterion. The one-one analysis simply addressed the question of whether children used as many different tag words as there were items in an array.

Table 8.1 summarizes the results of the analysis for the one-one principle in the spontaneous count sequences recorded in the [(2 vs. 3) − 1] and [(3 vs. 5) − 2] magic experiments. No effort is made to break down the results according to set size within a given experiment, because the children made errors on so few of the trials.

We consider first the results for the [(2 vs. 3) − 1] experiment, for which we have data for 2-year-olds as well as for 3- and 4-year-olds. Note that the younger the child, the greater the tendency to count. This observation is a replication of other findings (Beckmann, 1924; Descoeudres, 1921; Gelman, 1972b). What is noteworthy in this case is that the finding is extended to cover 2-year-olds.

Errors in one-one correspondence seldom occur. Even the 2-year-old's sequences are 80 percent error free. This is not to say that the youngest children counted as adults do. Many in fact did so, but five of the 2-year-olds used their own lists of tags (for example, one, thirteen, nineteen; two, six; one, four, three; *A, B*). Whatever tags they use, 2-year-olds show a remarkable tendency to say one number word for each item in an array.

TABLE 8.1. Use of the one-one principle in spontaneous counting.

Age group and experiment	*N*	Number of counters	Number of count sequences	Number of sequences with one-one errors	Errors by type			
					Partitioning	Tagging	Coordination	Other
2-year-olds								
[(2 vs. 3) − 1]	16	14	56	11	0	6	4	1
3-year-olds								
[(2 vs. 3) − 1]	32	25	69	9	0	0	8	1
[(3 vs. 5) − 2]	24	20	159	19	9	0	10	0
4-year-olds								
[(2 vs. 3) − 1]	32	16	25	0	0	0	0	0
[(3 vs. 5) − 2]	24	19	150	23	3	1	19	0

Not surprisingly, the older preschoolers make even fewer errors on sets of two and three items than the 2-year-olds do. Furthermore, the data from the [(3 vs. 5) − 2] experiment show that children as young as 3 years of age honor the one-one rule for sets of three and five items. For these set sizes, 88 percent of the 3-year-old count sequences and 82 percent of the 4-year-old count sequences met the one-one criteria. Thus we conclude that children of preschool age follow the one-one principle for counting—at least with respect to small sets.

Some errors that violated the one-one principle did occur. They fall into three categories, and the tendency to make errors of different types interacts with set size and age. Partitioning errors, that is, skipping an item or tagging it more than once, were observed in the (3 vs. 5) sequences and were made somewhat more often by 3-year-olds than by 4-year-olds. No such errors occurred in the (2 vs. 3) experiments. The effect of set size on this kind of one-one error seems straightforward. The more items the child has to count, the more likely it is that he will fail at some point to keep already counted items separate from to-be-counted items. Increasing set size still further should increase the tendency to make partitioning errors. We do not want to dwell too much on the increase in partitioning errors from set sizes of two and three to set sizes of three and five, however, for, the tendency to make such errors was slight. We conclude that children are able to conduct step-by-step partitioning of sets for the purpose of counting.

Occasionally a 2-year-old in the [(2 vs. 3) − 1] experiment made a tag-duplication error, assigning the same tag to different items (for example, one, four, four). The virtual absence of tagging errors in the 3- and 4-year-old sequences suggests that by the age of 3 children have learned enough count words to be able to give unique tags to as many as five items.

As Table 8.1 shows, the most common type of error in applying the one-one principle had to do with coordination. Such errors occurred in both the [(2 vs. 3) − 1] and the [(3 vs. 5) − 2] experiments at all age levels. They involved failing to count the last item in an array or using still another tag after all the items in the array had been tagged. In other words, a count of one too many or one too few was scored as a coordination error. One might argue that such errors are really partitioning errors. Perhaps the child misses the last item because he stops transferring to-be-counted items too soon. Perhaps the child who counts a nonexistent item is slow to recognize that he has transferred all the items to the already counted category. Given (*a*) that the one-too-many errors occur as frequently as the one-too-few errors, (*b*) that such errors occur across age groups and set sizes, and (*c*) that other partitioning errors do not necessarily accompany such errors, we are inclined to view these errors as a special class. They bring to mind the preschool children in various Russian studies (such as Luria, 1961), who did not know exactly when to stop a verbal accompaniment of a motor response. In other words, it is as if the children who make the one-too-many or one-too-few error have yet to perfect their ability to stop two coordinated processes at the same time. Their stop rules may be faulty. We know of no way to be sure of this hypothesis with regard to the data in Table 8.1. One reason we collected the videotape data that we present later in this chapter was to support our decision to classify the one-too-many and one-too-few errors as coordination errors rather than as partitioning errors.

The use of idiosyncratic lists was not ubiquitious. Such a tendency showed up only in the 2-year-olds, and even in this group number words were the prevalent items on the idiosyncratic lists. By 3 years of age, the children completely restricted their lists to number words and typically used them in the conventional order. Thus, nearly all of the evidence for the one-one principle involves the overt use of the English number words as tags. Yes, children watch "Sesame Street." But exposure to a given experience does not guarantee that, in Piagetian terms, it will be assimilated. We agree with Piaget that cognitive structures guide the organism's tendency to assimilate a given input from the environment. Therefore the fact that children assimilate the use of one-to-one correspondences between objects and number words leads us to postulate the one-one principle as a component of the cognitive structure underlying the development of counting behavior.

The one-one principle in the counting process can be stated as follows: In enumerating (counting) a set, one and only one numeron must be assigned to each item in the set. Since even our youngest subjects used count words as tags, we reformulate this principle as follows: For the child as young as 2½ years, enumeration already involves assigning one and only one numerlog to each item in the array. In other words, the child knows, although perhaps not consciously, that numerlogs are to be used as numerons.

Evidence for the stable-order principle. Counting involves more than the ability to assign arbitrary tags to the items in an array. Even if the child uses numerlogs to do this, he does not necessarily know the counting procedure. He must demonstrate the use of at least one additional principle—the stable-order principle. That is, there must be evidence that he uses tags (numerons) in a repeatable order.

To assess the degree of adherence to this principle, we analyzed the same protocols that we analyzed in connection with the one-one principle. We judged a child to be following the stable-order principle if when enumerating objects he (*a*) used the conventional sequence of count words, (*b*) used another standard sequence of terms (such as the alphabet), or (*c*) *consistently* used one nonstandard sequence. To meet this last criterion, a child had to count at least twice and to use the same nonstandard sequence across trials. The following protocol contains an example.

D. S., age 2 years, 6 months, during the surprise phase saw a rearranged three-item array as well as the original two-item array from Phase I. Eventually, the experimenter pointed to the two-item array and asked D. S. about quantity.

> How many on this [the two-item] plate? *Um-m, one, two.* How many on this [the three-item] plate? *One, two, six!* You want to do that again? *Ya, one, two, six!* Oh! Is that how many were on at the beginning of the game? *Ya.*

Table 8.2 summarizes the extent to which data from the (2 vs. 3) and (3 vs. 5) experiments can be taken to show the working of a stable-order principle in the control of the counting procedure. The table

TABLE 8.2. Use of the stable-order principle in spontaneous counting.

			Category of stable-order score (number of subjects)		
Age group and experiment	*N*	Counters	Always use stable order	Are shaky on order[a]	Show no stable order
2-year-olds					
[(2 vs. 3) − 1]	16	14	9	2	3
3-year-olds					
[(2 vs. 3) − 1]	32	25	23	1	1
[(3 vs. 5) − 2]	24	20	17	3	0
4-year-olds					
[(2 vs. 3) − 1]	32	16	16	0	0
[(3 vs. 5) − 2]	24	19	19	0	0

a. Correct on at least 60 percent of trials.

shows the number of subjects who counted, the number who always met the stable-order criteria, and the number who met the stable-order criteria on at least 60 percent of their count sequences. Subjects in this last group are credited with a *shaky* execution of the principle. The table also shows the number of children who gave no evidence of following a stable-order rule. We found very few children of any age who could not be credited with some inclination to use the stable-order principle. Of the 14 2-year-old children whose protocols were analyzed, 11 showed at least some grasp of this principle. And the 3- and 4-year-olds continued to honor the principle as set sizes increased up to five.

Readers may begin to wonder if we find *any* serious errors in the counting behavior of young children. They should remember that when we grant a counting principle—in this case the stable-order principle—we do not assume the flawless use of that principle. Our basic evidence for granting the stable-order principle comes from looking at the conditions under which very young children can accurately estimate number. Most very young children (2 to 3 years of age) can accurately estimate only very small sets (sets of no more than two to five items). These set sizes correspond closely to the range within which children use a sequence of numerlogs with a stable order. But we know that larger set sizes present difficulty for young children. Their accuracy at abstracting numerosity falls off as the set size increases. It is possible that their ability to follow the stable-order principle likewise falls off. As we will see later in the chapter, their use of this principle falls off only to a limited degree.

It is possible to argue that the children whose behavior is represented in Table 8.2 were not following a stable-order principle but were just repeating what they had had occasion to memorize. Several features of the data militate against this argument. For one thing, children did not have to use the conventional numerlog list to be scored as having the stable-order principle. We are particularly impressed with the occasional child who uses the alphabet in place of the conventional number-word sequence. The spontaneous use of a non-numerical sequence suggests to us that young children have a cognitive principle in search of an appropriate list, a principle ready to assimilate whatever usable list becomes available. It is the 2-year-old who uses the alphabet when counting. Does anyone seriously maintain that adult speech to 2-year-olds includes such forms as "Now, Billy, I want you to (*a*) pick up your toys, (*b*) find your storybook, and (*c*) come and sit on my lap." Gelman has never encountered such an example in her protocols of speech between adults and 2-year-olds (Gelman and Shatz, 1977)—even where the adults are academics. To us, the spontaneous use of the alphabet emphasizes the extent to which the child's behavior is guided by certain principles rather than by situation-specific rote learning. Although rote learning is necessary before the children can use the standard numerlog sequence, we suggest that children fasten on it precisely because it conforms to the stable-order principle of counting. How else to explain the idiosyncratic but stable lists of number words?

Anyone who watches young children count will surely notice the following features. Children practice counting toys, steps, cracks, or whatever—without being told to do so. Children appear to be aware that their counting sequences are unstable or unorthodox; evidence of this awareness is their tendency to correct themselves. Witness the behaviors of M. F. and A. B.

M. F., age 4 years, 5 months, participated in the [(3 vs. 5) − 2] experiment. The following exchange occurred in Phase I.

> Why does that [the three-item plate] lose? *Cause it's one, two, three.* Why does that [the five-item plate] win? *Because one, two, three, four, five, six. Mm-m. One, two, three, four, five.* [M. F. had counted the last item twice and then, without any input from the experimenter, recounted correctly. She immediately proceeded to count again.] *One, two, three, four. No, one, two, three, four, five. Cause it's five.*

A. B., age 3 years, 6 months, participated in an undergraduate's project on language acquisition. During the course of a long interview, she encountered a display of eight items and began counting.

> *One, two, three, four, eight, ten, eleben. No, try dat again. One, two, three, four, five, ten, eleben. No, try dat again. One! two! three-ee-four, five, ten, eleben. No.* [This pattern of self-correction continued for many attempts and ended with the following count, which may or may not have been error free.] *One, two, three, four, five, six, seven, eleven! Whew!*

Such rehearsals are characteristic of early language-learners—learners who are assumed to be practicing a rule just learned (Weir, 1962). They are also characteristic of the sensorimotor patterns of Jacqueline, Lucienne, and Laurent, Piaget's children. Developmental psychologists consider these examples to be instances of knowledge structures that guide the assimilation of environmental stimuli. We see no reason not to claim similar examples as evidence of the presence of the stable-order principle.

Thus, we interpret the data in Table 8.2 as evidence of an underlying counting principle. We of course do not claim that children have innate knowledge of the number-word sequence. This list obviously needs to be learned; but it begins to appear that the learning is facilitated and guided by an already present cognitive principle.

We have stated the stable-order principle as follows: Numerons used in counting must be used in the same order in any one count as in any other count. Given the data presented in Table 8.2, we can reformulate the principle to say the following: For the child as young as 2½ years, enumeration already involves assigning numerlogs in the same order in any one count sequence as in any other count sequence, at least for sets of two to five items.

Evidence for the cardinal principle. Counting involves more than the unique tagging of items and the use of a repeatable list of tags. Because of the special status assigned to the final tag in a given tagging sequence, counting allows one to determine how many items a set contains. The final tag, unlike any of the preceeding tags, is taken to represent a property of a set as a whole, namely its cardinal number. This role of counting in the definition of numerosity leads us to consider the cardinal principle a component of the counting process.

Before we go on, it is important to make clear what we are and are not doing by incorporating the cardinal principle into our definition of the counting procedure. We know that the counting procedure can yield a representation of the cardinal numerosity of the set. We also assume that a young child who arrives at such a representation may be using this procedure. Indeed whenever the evidence indicates that a child has singled out the last tag in a particular count sequence, we say that he has used the cardinal principle. We do not assume that counting is the only way individuals might arrive at a representation of the cardinal numerosity of a set. Nor do we assume that a child who uses the counting procedure to establish such a representation has a full appreciation of all the properties of cardinal number.

There are several set-theoretic definitions of cardinal number, which we will discuss in Chapter 11. All these definitions involve the application of the one-one principle in the comparison of more than one set. In general, when the elements of two different sets can be placed in one-to-one correspondence, the two sets are said to have the same cardinal number. Other sets that can be placed in one-to-one correspondence with these sets also have the same cardinal number.

Note that here the definition of cardinal number is bound to the definition of equivalence. For the moment, we remain neutral on the issue of whether the definition of equivalence of sets of different objects and the definition of cardinal number are so intricately tied together. We simply point out that it *is* possible to ask "How many things are here?" in reference to a single set. And the counting procedure is well suited to answering such a question.

We alluded above to the behavioral criteria by which we infer the use of the cardinal principle, namely an indication that the child treated the last tag in a sequence differently from earlier tags in that sequence. Specifically, we judged a child to have followed the cardinal principle if he met at least one of four criteria. Two of these criteria involve behavior that we take to provide *direct* evidence for the use of the cardinal principle. To meet the first of these direct criteria, the child counted (correctly or incorrectly) the number of items in the array and repeated the last tag: for example, "two, sixteen; sixteen"; "one, two, three; three!"

As in this second example, the repetition of the last tag was sometimes given emphatic stress. This led us to consider whether children sometimes used exaggerated stress for the final word in a count without repeating that word. Finding that some did, we decided to include this stress as our second direct criterion. We recognize that the children we observed use the intonation patterns of English and that therefore some stress is to be expected on the final word of many sequences. In these count sequences, however, the preschoolers often exaggerate the stress on the final word well beyond the bounds of ordinary intonation patterns, giving the impression of a shout or a scream. Only such exaggerated stresses were counted as evidence of the cardinal principle according to the second criterion. The transcribers of the protocols were not primed to look for stress; presumably they noted only those cases that were exaggerated. In any case, for now we include this second criterion, keeping in mind the need to obtain evidence of its validity before using it in further studies.

It is interesting to note that Bereiter (1968, p. 14) advises teachers of young children that "the last number counted should be given special emphasis, both with voice and touch, following which the adult should present a statement and question on this order: Five crayons! . . . How many crayons?" Such a teaching method incorporates both of our direct criteria for evidence about the cardinal principle. We infer that Bereiter, too, has observed young children using both criteria.

We also judged the presence of the cardinal principle by two *indirect* criteria, both of which concerned number words that were not immediately preceded by count sequences. A child met the first of these criteria if he assigned numerlogs to each of the items in the set on one trial, for example "one, two, three," and shifted to a simple indication of numerosity of the same set without overt counting on a later trial. The children could meet this criterion by using their own idiosyncratic lists of numerlogs. Thus if a child said "one, two, six" on first encountering a three-item array and later said "six" to describe the same array, he was scored as having used the cardinal principle by this criterion.

Children met the second indirect criterion if they assigned the correct value to the set without counting aloud. We did not score children as having used the cardinal principle if they simply stated an incorrect value. To do so would amount to claiming that the child who says numbers words out of the blue understands the cardinal principle.

In a way, by using these two indirect criteria we are reifying our hypothesis that correct estimates of numerosity are almost always based on counting, whether covert or overt. Critics might object that stating the correct numerical value of a set should not be taken as evidence for the cardinal principle of counting since the estimate may not have been obtained by counting. It could be argued that the child simply forms an association between the perception of, say, three objects and the word *three.* As is evident from Table 8.3, the ability to simply state the numerosity of a given set without overt counting is a later development, suggesting to us that it is based on the development of covert counting skills (compare Gelman and Tucker, 1975). However, in our presentation of the data, we keep separate the direct and indirect criteria, so that the reader who is skeptical may draw his own conclusions.

Table 8.3 shows the number of children who met the various cardinal criteria, either separately or in combination, at least once. We had some difficulty with the question of how many times we should require a child to show this principle. The basic problem is that we were not asking our subjects any particular question that might elicit the use of this principle. Furthermore, the criterion that requires counting first and simply labeling later seems unlikely to be met more than once. For these reasons we decided to use an at-least-once rule for the cardinal principle.

As Table 8.3 shows, the cardinal principle is clearly used by chil-

TABLE 8.3. Use of the cardinal principle in spontaneous counting.

Age group and experiment	*N*	Number who stated a cardinal numerosity for set	Cardinal criteria (number of subjects)			
			Direct	Indirect	Both	Neither
2-year-olds						
[(2 vs. 3) − 1]	16	16	5	5	4	2
3-year-olds						
[(2 vs. 3) − 1]	32	31	5	10	16	0
[(3 vs. 5) − 2]	24	24	4	5	15	0
4-year-olds						
[(2 vs. 3) − 1]	32	29	0	21	5	3
[(3 vs. 5) − 2]	24	22	2	5	15	0

For this analysis we include subjects who did not count but did use a numerlog to represent the numerosity of the set. In Tables 8.1 and 8.2 only subjects who counted aloud could be classified. Children who never referred to number are excluded from this analysis.

dren as young as 2½. Note that indirect indications of the use of cardinal principle increase with age: 64 percent of the 2-year-olds, 83 percent of the 3-year-olds, and 100 percent of the 4-year-olds who applied the cardinal principle met the indirect criteria.

One fact that Table 8.3 does not reveal is that only the youngest subjects were credited with the cardinal principle solely on the basis of the stress criterion. This fact could be taken to mean that we should drop this criterion, in which case two of our youngest subjects would move to the "Neither" colmn of the table. Alternatively, it could be taken to mean that the stress criterion indexes the beginning stages of the acquisition of the cardinal principle. If this is true, we should see children using stress patterns to index cardinal number at about the time they master a given set size. Accordingly, we might expect older children to satisfy the indirect criteria for set sizes they have already mastered and the direct criteria for set sizes they are just mastering. Further, a close inspection of carefully collected count sequences might reveal the restricted use of the stress criterion for set sizes just within the range of the child's competence. Even if we decide to exclude the stress criterion, it remains true that three-fourths of our youngest subjects show evidence of possessing the cardinal principle.

The data from the [(3 vs. 5) − 2] experiment also indicate that 3-

and 4-year-olds show clear evidence of being able to apply the cardinal principle. Very few children in these age groups met only the direct criteria. Of those who did, there were none who met only the stress criterion. This result is perhaps unfortunate, for it leaves us unable to resolve the question of the validity of the stress criterion as an index of the emergence of the cardinal principle. We will have to wait until we consider how children of this age perform with still larger set sizes.

On the basis of the results in Table 8.3, we conclude that the cardinal principle is present even in the youngest children we have studied, 2½-year-olds. We also find that the ability to give the last tag in a sequence a special status is manifested in diverse ways, suggesting that full competence with this principle may emerge by passing through a number of steps.

The cardinal principle may be stated as follows: The final numeron assigned to the last object in the set represents a property of the set—its cardinal number. Our findings allow us to reformulate this principle as it applies to young children, as follows: For the child as young as 2½ years, enumeration already involves the realization that the last numerlog in a set—at least in a small set—represents the cardinal number of the set.

Evidence for the combined use of the how-to-count principles. Having analyzed the magic experiment protocols from the standpoint of each of the three principles in the child's counting process, we now examine how well the children combined these three principles into an integrated counting procedure. Table 8.4 presents the percentage of children in each age group who manifested each possible combination of principles. Considering first the results of the [(2 vs. 3) − 1] experiment, we find that approximately 40 percent of the children at each age level received perfect scores on all three principles. A closer consideration of the results in Table 8.4 leads us to conclude that the vast majority of children utilize or try to utilize all three principles.

First, there are the children who used all three principles but not consistently across count trials. We gave such children credit for *shaky* mastery of the principles. We feel justified in adding these children, at the very least, to the pool of those who showed evidence of using all three principles. But including only these children and those with perfect scores would yield a peculiar result, namely, that 2- and 3-year-olds are better at counting than are 4-year-olds. Only 41 percent of the 4-year-olds, as opposed to 63 percent of the 2-year-olds and 72 percent of the 3-year-olds, fall in these two groups. Note that in the

TABLE 8.4. Combinations of the three how-to-count principles in spontaneous counting.

Age group and experiment	N	Combination used (percentage of subjects)					
		All three perfect	All three, one shaky	One-one and stable-order	Stable-order and cardinal	Cardinal alone	No reference to number
2-year-olds							
[(2 vs. 3) – 1]	16	38	25	13	0	25	0
3-year-olds							
[(2 vs. 3) – 1]	32	38	34	3	3	19	3
[(3 vs. 5) – 2]	24	33	50	0	0	17	0
4-year-olds							
[(2 vs. 3) – 1]	32	41	0	9	0	41	9
[(3 vs. 5) – 2]	24	54	21	4	0	17	4

[(2 vs. 3) – 1] experiment no 4-year-olds appear in the "all three but at least one shaky" category. If they show evidence for all three principles, they do so unfailingly. But if 4-year-olds are so good at following the principles, why are only 41 percent scored as having all three? We think this puzzling result is an artifact of our scoring criteria. Some children received credit for showing only the cardinal principle. These children did not count aloud but simply stated the numerlogs that represented the numerical value of the arrays. Since they did not count overtly, we could not score them in the one-one and stable-order analyses. This group included 41 percent of our 4-year-old subjects. Should we assume that they lacked the ability to use the one-one and stable-order principles? We think not. For one thing, almost all of the 4-your-olds correctly counted their prizes at the end of Phase I, and none had received fewer than five prizes. So they counted larger set sizes than those being discussed here. Why then, did they not count the smaller arrays? Recall that 4-year-olds are less inclined than younger children to count small set sizes overtly. Only with larger set sizes are 4-year-olds consistently observed to count aloud. These facts lead us to conclude that children who simply label, rather than counting, set sizes of two and three have progressed beyond the need to count overtly in order to represent such small numerosities. Accordingly, we suggest that they be considered to possess all three counting principles. If we grant them all three principles, we find that 88 percent of the 2-year-olds, 91 percent of the 3-year-olds, and 82 percent of the 4-year-olds are able to successfully coordinate the three how-to-count principles when faced with arrays of two or three items. The reader who is disturbed by our assumption that children who demonstrate only the cardinal principle also possess the one-one and stable-order principles will be reassured by our later finding that children who label small sets count larger ones correctly.

The fact that so many children can be credited with having all three principles makes it difficult to investigate the relative difficulty of the principles. The percentages in the remaining cells are small but do give us a hint. Children in these cells are most likely to use the combination of the one-one and stable-order principles. That is, they are most likely to show evidence of being able to count the number of items in an array without being able to assign a cardinal value to that array. This evidence suggests that the ability to apply the cardinal principle lags behind the ability to apply the one-one and stable-order principles. We will return to this issue when considering videotape data for children counting a wider range of set sizes.

The scores from the [(3 vs. 5) − 2] experiment also indicate that children are able to use the three principles in concert. The 3- and 4-year-old subjects differ in one notable way: The 3-year-olds are more likely to be shaky in their use of at least one of the three principles.

A comparison of the results of the two experiments (see Table 8.4) is informative in two ways. First, 4-year-olds count aloud more with the larger set sizes than with the smaller set sizes. This observation is reflected in the fact that very few 4-year-olds demonstrate only the cardinal principle with the larger sets. Second, the [(3 vs. 5) − 2] experiment brings out some shaky counters in the 4-year-old group. Thus, as expected, set size increases both the tendency to count aloud and the tendency to err while doing so.

Summary of the Magic Study Analyses

Our analysis of the data from Gelman's magic experiments shows that young children follow the three basic how-to-count principles of our counting model—at least when dealing with homogeneous, linearly arranged sets of no more than five items. Children make relatively few errors on the various principles. The one-one errors that occur tend to be either partitioning errors or coordination errors. Coordination errors (overcounting or undercounting) occur for all set sizes; partitioning errors (skipping or double counting) seem to be restricted to the larger of the set sizes we used. Tagging errors (using the same tag more than once) are almost nonexistent. Together these findings suggest the following conclusions: that preschoolers have available a list of tags long enough to allow them to follow the one-one principle; that they have some difficulty coordinating the selection of tags with the step-by-step partitioning of items (thus the tendency to select one too few or one too many tags); and that increasing the set size increases their difficulty in keeping track of the difference between counted and to-be-counted items. If these interpretations are correct, what might we expect when children confront still larger set sizes? We predict that partitioning errors will become more common and that coordination errors will continue to appear. Together these errors may initially seem to account for children's inability to count larger set sizes. On closer inspection, however, we may find that their unstable ability to coordinate the components of the one-one principle accounts for more of the difficulty. We also predict that tagging errors will begin to appear. As we have already suggested, one major difficulty children will confront when counting larger arrays will be the need to have as many unique tags as there are items to tag. Larger sets should make this straightforward problem of list learning more prominent. It will be interesting to monitor the way children deal with this problem. They may repeat tags, "one, two, three, four, four"; recycle a list, "one, two, six, eight, one, two, six, eight"; introduce illegal tags, "one, two, three, cow"; or even limit their count to as many items as they have tags. We make no predictions here but simply point to the variety of interesting errors they might make in trying to apply the one-one principle in the face of the limitations of their stably ordered lists of tags.

It is worthwhile to look more closely at the question of what constitutes a coordination error. We have scored a coordination error when children assigned one too many or one too few tags to the items in an array. Upon reflection, we see that this is a gross index. We have no way of telling whether a child who made such an error omitted a tag for (or double-counted) the first or the last item in the array. At the beginning of a count such an error would indicate trouble in starting the process; at the end of a count it would signal trouble in stopping the process. A difference clearly exists between starting and stopping a complex motor-verbal sequence. The evidence suggests that the latter is generally more difficult (Luria, 1961). We would feel more at ease with our decision to classify these errors as coordination errors if we found that in counting, too, it is harder for the young child to stop than start. This is not to say that we should find no starting errors. Indeed, if coordinating the components of the one-one principle is a problem for young children, they should make *some* errors at the beginning of a count. Since we cannot retrieve the relevant information from audiotape transcripts, the answers to these questions must wait for the analysis of our videotape data.

We have other reasons for moving to videotape. Anyone who watches young children count notices the ubiquitous use of pointing. We have mentioned how pointing might aid the child in the coordination of the tagging and partitioning components of the one-one principle. But by choosing to point, the child further complicates the task before him. Now he has to coordinate pointing with the other components. He may miss an object that he intends to point to, point to an item more than once or never, point to a space, and so on. Furthermore, he has to start and stop the pointing procedure at the same time as he starts and stops the tagging and partitioning procedures. (Note that we are still talking only about the one-one principle. And some think that counting is a trivial skill!) At any rate, to study the role of pointing, we obviously need videotape data. Such considerations led to Gelman's decision to videotape all further data collections concerning counting.

In our analysis of the use of the stable-order principle, we noted that our older subjects typically used the conventional number-word list but that the 2-year-olds showed some tendency to use idiosyncratic lists—even for set sizes of two and three. We anticipate that when older children are asked to deal with larger set sizes they will use lists that begin with the conventional sequence but shift over to idiosyncratic sequences, such as "one, two, three, four, five, eight, ten, nineteen."

Three findings from the cardinal principle analyses stand out: First, older subjects give mostly indirect evidence of the use of the cardinal principle, stating the numerlog that represents the cardinal numerosity of the set without counting out loud to get it. If there is indeed a developmental tendency to shift from tagging each item with a numerlog and repeating or emphasizing the last one to simply stating the last one, we should find a similar trend within an age group as set size increases. That is, a child who gives only indirect evidence of using the cardinal principle in counting sets of five or fewer items might give more direct evidence when dealing with a set of seven items. Second, we credited children with possession of the cardinal principle if they placed unusual stress on the last tag in their list. The tendency to do this was more prevalent in the youngest children, suggesting that it is an early index of an emerging understanding of the special status of the last tag in a count list. It is also possible that this stress is the easiest way for the child to signal his knowledge that the last tag should be treated differently. Before we continue to include this criterion in scoring for the use of the cardinal principle we should look for evidence that it is used by older children. Again, research with larger set sizes is required. If the use of stress is an early developmental index of the child's understanding of cardinal number, then older children might return to it when counting larger set sizes.

Finally, the tables contain a hint that the child's ability to apply the cardinal principle emerges somewhat later than his ability to apply the one-one and stable-order principles. It is as if the child needs to practice tagging a given set size before he can focus on the last tag on the list. Such a line of argument suggests that children should be less likely to meet the criteria for the cardinal principle when counting larger sets. In other words, when we introduce larger sets we should find more subjects than we have found so far who use the one-one and stable-order principles without the cardinal principle.

The Videotape Counting Study

Issues of Methodology

Subjects. The children who participated in this experiment attended a YMCA day-care center in Center City, Philadelphia. The children at the center tend to come from white middle-class homes, but the sample represents a reasonably diverse economic and racial population. Before we started our project, we obtained permission from parents. All parents agreed to let their children participate, and we worked with all children in the school: 19 2-year-olds, 21 3-year-olds, 19 4-year-olds, and 15 5-year-olds.

Design. We tried—not always successfully—to test all the children on set sizes 2, 3, 4, 5, 7, 9, 11, and 19. We chose these set sizes for two reasons: to include small and large sizes; and to use set sizes that we had worked with in other experiments.

The design called for giving each child six trials with each set size of 3 or more—three trials with linear arrays and three with nonlinear arrays. Since a set size of 2 can only be arranged linearly, there were but three trials for it. All the tests in the magic experiments were conducted with linear arrays. Given our component analysis of the one-one principle, we might expect children to have more difficulty in keeping track of items arranged haphazardly. Another difference from the magic experiments is that here we used arrays that were heterogeneous with respect to color. This was done to see if our previous results generalized to heterogeneous materials.

The reader familiar with the difficulties of keeping young children task-oriented will worry about the demands our design made. We were asking children to sit through 45 trials of what might be quite a boring task. To mitigate the anticipated boredom, we introduced several features into the design. First, we blocked the set sizes into pairs of 2 and 3, 4 and 5, 7 and 9, and 11 and 19. Each pair of set sizes was displayed on a pair of lazy Susans, so that the arrays were easily rotated. The child was told that one plate was for him, the other for the experimenter or her helper, a puppet. The puppet was used because we had previously found that young children are much more inclined to answer to and do things for a puppet; furthermore, the puppet's presence increases attention span and tolerance for a lengthy testing procedure.[1] Third, we ran the trials in an easy-to-hard sequence. The trials involving set sizes 2 through 5 were run before the trials involving the larger set sizes. All trials for a given pair of set sizes were run in rapid succession.

We are convinced that we took as many steps as possible to gain the cooperation of our subjects. But even the best-laid plans regarding work with preschoolers go astray. Therefore, we decided in advance that the child's proclivities would take precedence over design considerations. If a child wanted to keep working with a given set size, he was allowed to do so; and if it was clear that working with larger set sizes was frustrating the child, the experimental session would stop.

Procedure. The experimenter spent several hours a day for at least one week in each of the classrooms in the school to become known to the children. Then she took each child separately to the room in which we were to conduct the research and simply played with him, using toys appropriate to his age. Finally, she asked the child to participate in the experiment proper. Altogether the experimental session lasted approximately 15 minutes.

When the child entered the room for the experiment, he was asked to sit at a small table. The video equipment was hidden behind a set of room dividers, one of which had a small opening for filming.

To start the counting experiment, the child was shown the two lazy Susans and told that one of the plates was for him, the other for the puppet. Then the designated pairs of set sizes were constructed by randomly drawing the appropriate number of items from a container and placing them in position (either linearly or nonlinearly). The length and shape of arrays were varied haphazardly to avoid producing systematic perceptual cues for a given set size.

To begin the testing, the experimenter (in either her own voice or the high-pitched, squeaky one of the puppet—whichever seemed best suited to the child) asked the child to indicate "how many" items were on his (or the puppet's) plate. Questions asked in the trials for a given set size varied: "how many," "count them," "now how many," "go ahead." With one exception, no attempt was made to vary the questions systematically. The exception was that the experimenter always started by asking "how many?" Here, as in previous work, we found that children typically counted. Some simply gave a number as the answer. These children were asked to count on a subsequent trial to

1. Merry Bullock deserves the credit for this serendiptitous discovery of how to gain the cooperation of young children.

allow us to determine whether they were able to count. As we will see, some of the older children did not count arrays representing small set sizes even when asked.

After an initial presentation, the arrays were rotated or rearranged. If a child's attention appeared to be drifting away from a given array, the experimenter removed the display and substituted another of the same numerosity rather than rotating the array.

Videotape data reduction process. In transcribing a given session, the transcriber followed a combination of scoring, coding, and protocol-copying instructions. Thus the term *transcript* is something of a misnomer. Our goal was to produce as detailed a description of a trial as possible without being forced to record verbatim everything the child said and did.

The videotaped sessions were transcribed trial by trial. Initially the transcriber indicated for each trial what set size was used, how the array was arranged, what questions the experimenter asked, and whether the child treated the trial as a count trial or a cardinal-only trial. A count trial was indicated if the child gave any evidence of tagging separate items. A trial was defined as a cardinal-only trial if the child simply answered a question with one numerlog, as in "There's three there." Any false start on a count trial, such as "one, three; no, one, two, three," was noted, although subsequent analyses of that trial excluded the false start.

If a child treated a trial as a cardinal-only trial and erred, his answer was written into the transcript. There were no further transcription instructions for such trials, but the transcriber used a column labeled "other comments" to indicate any relevant aspects of the child's behavior (for example, if the child held up a given number of fingers or was not looking at the array when he answered).

Having separated the count trials from the cardinal-only trials, the transcriber then separated the perfect count trials from the imperfect ones. The child who said "one, two, three, four, five" in response to questions about a five-item array was scored as perfect on that trial.

On perfect count trials, the transcriber made note of the child's pointing behavior, indicating whether the child pointed, whether he pointed to each item he tagged, whether he started to point when he started to tag, and whether he stopped pointing when he stopped tagging. Any other observations about pointing behavior (such as that the child pointed into the air) were entered in a "comments on pointing" column. Any other notable behaviors (for example, if a child paused in the middle of a count and then continued) were entered in the above-mentioned "other comments" column. In a separate column, the transcriber noted the pace at which a child counted as fast, neutral, or slow. Finally, the transcriber noted whether the child repeated or stressed the last tag.

The transcription of imperfect counts was much more elaborate because we wanted to characterize the imperfections in as much detail as possible.

If the child did not use a conventional list of numerlogs, the ones he did use were transcribed. If he used as many separate tags from an idiosyncratic list as there were items to be counted, the trial was subsequently coded as if he had correctly used a conventional list. If he did not use the same number of tags as there were items, or if he repeated one or more tags, the trial was coded in the following manner: The tags he used were written down. If he made a tagging error, this was entered and scored as either a repeat error or a label error. A repeat score indicated that the child repeated a tag, as in "one, two, two, four"; a label score indicated that the child used some unsuitable numerlog, as in "one, two, blue," or "one, two, three, a cat." If the child made a partitioning error, its exact location was noted and it was scored as either an omit error or a double-count error. An omit error was scored if the child failed to tag an item in the middle of an array. A double-count error was scored if the child tagged an item in the middle of an array more than once. Which item was so tagged was recorded in the "other comments" column. Omitting a tag on the last item in an array was scored separately, as a coordination error rather than a partitioning error. So was double-counting the first or last item in an array. Careful study of the videotapes convinced us that a child who double-counted an initial item had difficulty starting the counting procedure and that a child who double-counted a final item had difficulty ending the counting procedure. Therefore, we felt justified in treating these kinds of behavior as coordination errors.

Coordination errors also included overruns—sequences in which the child went on chanting tags after he came to the end of the to-be-counted items. Overruns differed from double counts of the last item in that the additional tags did not appear to be assigned to the last (or any other) item. The extent of the overrun was noted. Finally, we scored a general asynchrony as a type of coordination error: in such cases, the children seemed to get caught up in the production of tags and forget to systematically assign these to anything, be it a point, an object, or both. They were like the conductor who allows his baton to take over, but in this case the baton was in the vocal tract.

Pointing behavior on imperfect count trials was transcribed like that observed on perfect trials: The features recorded included whether the child pointed, whether he started and stopped pointing when he started and stopped tagging, and whether he pointed to each item, to spaces, to the air, and so forth. As in the scoring of error-free trials, the pace of the child's counting was noted. Finally, the transcriber noted in the "other comments" column any further information that seemed worth transcribing.

In all, the transcriber worked with 30 coding or comment columns per trial. It quickly became apparent that the above instructions were ill-suited to the transcription of the sessions with 2-year-olds. Otherwise, the transcriber reported that she was able to work with the instructions. Both of these impressions were confirmed by a check of a random 25 percent of the transcripts. One of two independent raters watched the videotape and went through the transcripts, keeping track of agreements and disagreements regarding coded data as well as clear recording errors (such as omitting a trial, indicating an array was displayed linearly when it was not, or coding an error in the wrong column). The raters agreed with the transcriber regarding codes and what was recorded on the transcripts on 84 percent of the trials. If we exclude the two 2-year-olds' transcripts, then they agreed on 89 percent of the trials. Excluding obvious errors in recording brings the percentage of agreement to 88 percent for the sample of all age groups and 97 percent for the sample without the 2-year-olds. The fact that including the 2-year-olds raised the disagreement rate supported our impression that the transcription instructions would not work in reducing the youngest children's data and led to our decision to set aside the 2-year-olds' tapes for separate consideration. It was clear, however, that we could use the transcription procedure for all other age groups and count on obtaining a high degree of accuracy in the transfer of videotape information to transcripts. The tapes themselves were always available, of course, in case one found the transcription inadequate or mistrusted it.

Results For 3-, 4-, and 5-Year-Olds

Preliminary findings. The crudest index of emerging counting skills is whether or not children use as many counting tags (unique or not) as there are items to be counted. This they generally do. Figure 8.1 summarizes the extent to which children use either N or $N \pm 1$ tags for a set with N items. It is important to note that these curves represent only those children who did try to count a given set size. We did

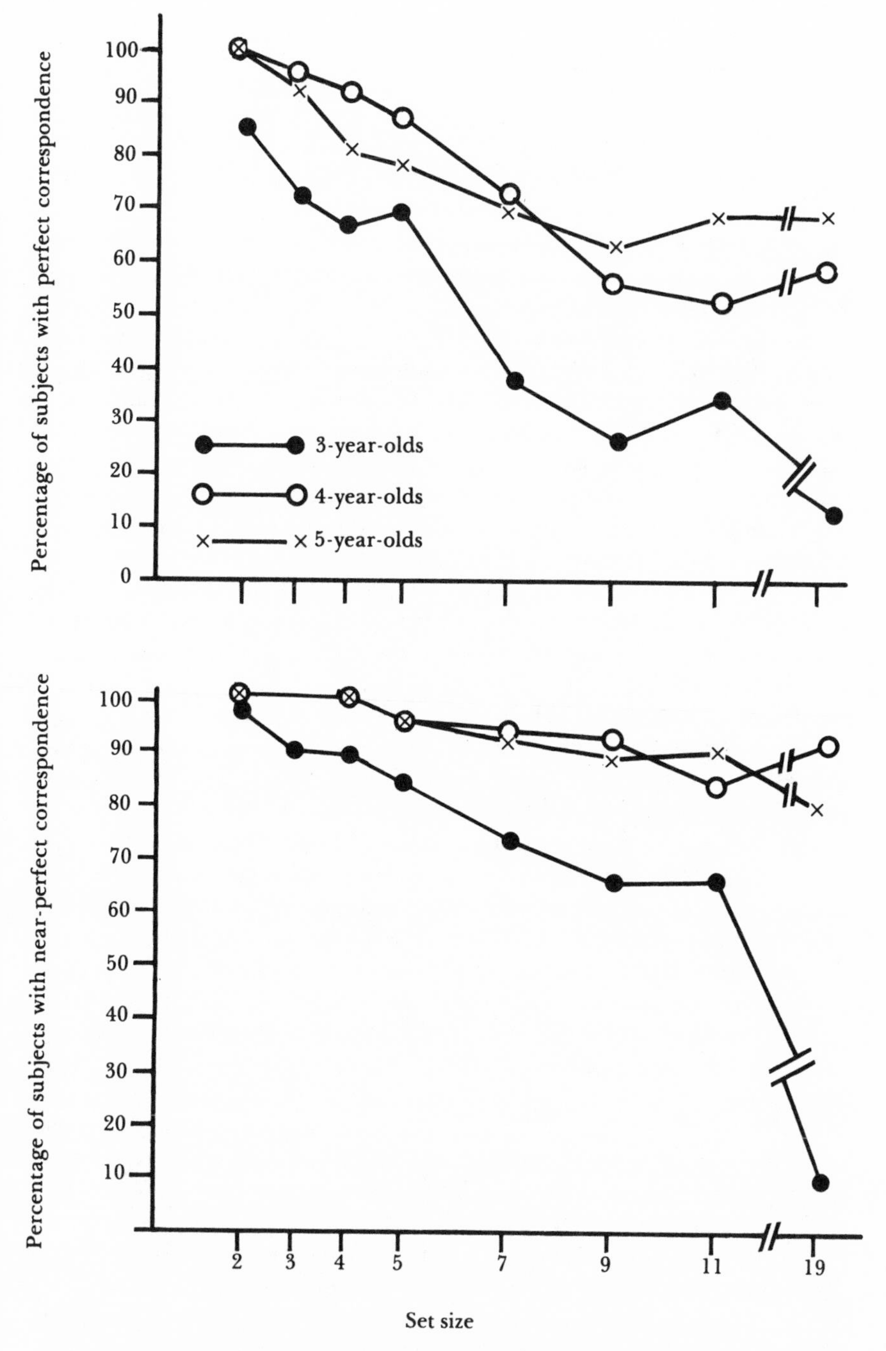

Figure 8.1. Percentage of children who achieved perfect or nearly perfect correspondence between set size and number of tags used, shown as a function of set size. Upper graph shows those who used exactly N tags for a set of N items; lower graph also includes those who used $N \pm 1$ tags.

not assign a score of zero to those children who refused to continue in the experiment. As will become clear in later analyses, no more than 60 percent of the 3-year-olds are represented for set sizes 9, 11, and 19 in Figure 8.1. The 4- and 5-year-olds tended to stay in the experiment no matter what the set size.

In the analyses represented in Figure 8.1, the tags used by a given child did not have to be unique but did have to meet our definition of a numerlog. Any number word or letter of the alphabet was counted as a tag. (In fact, children used only number words—an observation that is congruent with the magic protocols, which showed only 2-year-olds using tags from the alphabet). In this analysis it did not matter if a given number word was used more than once in a counting sequence. A child who counted "one, two, two" for a three-item array was scored as having used as many tags as there were items in the array. This analysis, unlike the analysis of evidence for the one-one principle, simply sought to determine whether the children understood that counting requires tagging all items.

The top panel of Figure 8.1 summarizes, for the three age groups, the extent to which the number of tags used corresponded perfectly with set size. The bottom panel shows the number of children of each age group who used either the correct number of tags or one too many or one too few tags for a given set size. Our interest in the children who missed the correct answer by one tag derives from the hypothesis that such an error reflects a coordination problem more than a lack of understanding that the number of tags and number of items should be the same. As support for our intuition on this matter, we offer our own errors in counting out set sizes greater than 9. When we viewed the videotapes, we noticed that, on occasion, the number of items placed on an array deviated by one item from the intended N of 11 or 19!

Figure 8.1 supports our initial statement that the children understood that they were to assign as many tags as there were items in a set. As might be expected, set size affected the extent to which children met this requirement perfectly. Also, the older children were better able to meet this criterion, particularly for the larger set sizes. When the 4- and 5-year-olds failed to use as many tags as items, they missed by just one tag ($N \pm 1$) even for set size 19. The 3-year-olds did not do quite as well as the older children, but with the exception of set size 19, they did perform creditably on the larger sets. For set sizes 7, 9, and 11, 73 percent, 65 percent, and 67 percent of the 3-year-olds used N or $N \pm 1$ tags.

Note the drop below 70 percent for 3-year-olds on set sizes larger than 7. Recall that many 3-year-olds withdrew from the experiment at just this point. We find it interesting that some children pull out of an experiment at the point where others of their age start to falter. This fact raises the intriguing possibility that a young child's lack of cooperation may be related to a sense that he will not do well at the task.

At the very least, we can now conclude that when young children count they know they are supposed to assign as many tags as there are items in the array. What is more, they can do so for set sizes of at least 7, and indeed, 4- and 5-year-olds can do so for set sizes up to 19. These findings provide further evidence for a conclusion we reached in Chapter 5, namely that young children do *not* treat set sizes greater than 5 as undifferentiated *beaucoups.* To be sure, the evidence presented here is somewhat weaker than that in Chapter 5, but the agreement is worth noting nevertheless. As might be expected, plotting the present data in the format of Figure 5.1 yields results similar to those in Figure 5.1. Figure 8.2 shows the median number of tags assigned across all trials for a given set size and age group. The horizontal bar at the bottom of a range line indicates the number of tags above which 90 percent of the responses fell; the horizontal bar at the top of the range line indicates the number of tags below which 90 percent of the responses fell. Several points about these results are noteworthy. First, the median number of tags used increases systematically as a function of the size of the set being counted. For the 3-year-olds, the median number of tags used falls short of the number of items for set sizes of 9 and 19. Just as we found that the median judgment of numerosity for 4- and 5-year-old children matched the set size (Figure 5.1), we find that the median number of tags children of this age used in counting matches the set size. Second, the ranges accounting for the middle 80 percent of the answers given are remarkably narrow—except for 3-year-olds' answers for set size 19. Third, except for the 3-year-old group, the range in the number of tags used to count a set of 19 does not overlap with the range used to count a set of 11, and likewise for most other pairs (11 and 9, 9 and 7, and so on). For the 3-year-olds the ranges do overlap somewhat for the small set sizes. Not surprisingly, the children represented in Figure 8.2 do better than those represented in Figure 5.1. We might expect children to know that the number of tags should match the number of items before they know which tag is conventionally used to represent the numerosity of a given set.

In addition to the assigning of numerlogs, pointing behavior

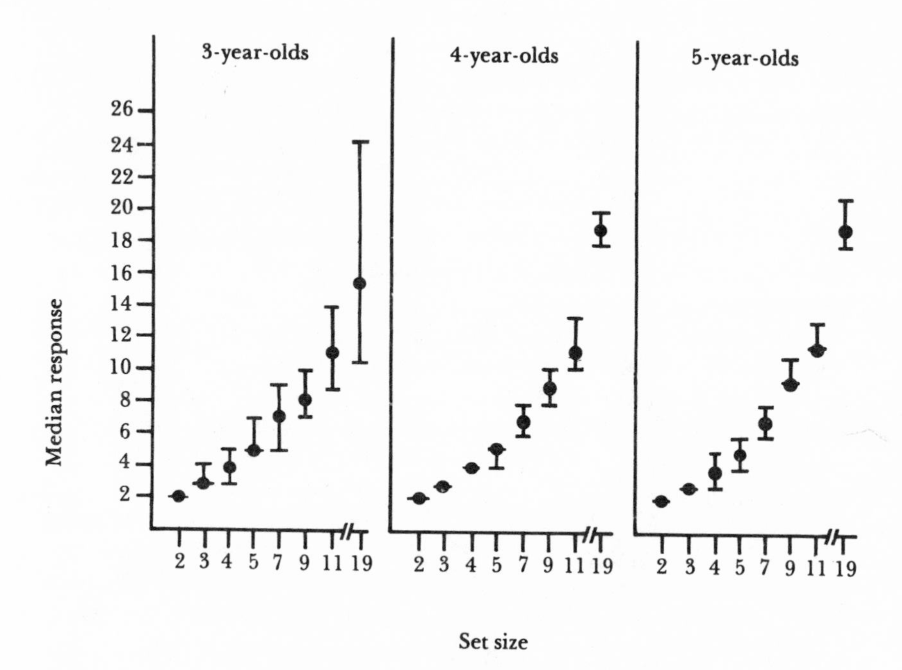

Figure 8.2. Median and range of the number of tags used in counting sets, as a function of set size. The indicated ranges include all but the highest 10 percent and the lowest 10 percent of the numbers of tags used.

seems to be central to the counting procedure. To our surprise, almost all children of all ages point when they count. We observed children pointing on a full 98 percent of the count trials. The children not only pointed but also were remarkably inclined to start and stop pointing as they started and stopped tagging. Even 3-year-olds started and stopped the two processes together on 92 percent of their count trials; 4- and 5-year-olds did likewise on 98 percent and 97 percent of their respective count trials. Note that these children did not necessarily point to every item or tag every item. But even if they were poor or sloppy counters, they at least knew how to coordinate the beginning and end of a count with the beginning and end of the use of pointing. Children probably pick up this skill from those who try to teach them to count. We do not mean to suggest that they do not. What impresses us is how common such behavior is. As we argued earlier, pointing is a handy device for coordinating the tagging and partitioning components of the one-one principle. We now turn to considering how well this device serves the children in their application of the one-one principle.

Evidence for the one-one principle. In our analysis of this principle we determined whether children assigned each to-be-counted item a *distinct* number word. For a given set size, a child could be scored as perfect, shaky, questionable, or no on this principle. To be scored as *perfect,* a child had to be correct on at least 90 percent of the trials for a given set size. For children who received the usual six trials, only a score of six out of six meets this criterion. However, some enthusiastic subjects insisted on more than six trials. (Recall our discussion of spontaneous counting earlier in this chapter.) Children who were correct on between 60 percent and 89 percent of their count sequences for the given set size were scored as *shaky.* Children who were sometimes correct but failed to meet the 60 percent criterion were scored as *questionable.* The *no* category was used for children who erred on every trial.

A trial was scored as correct only if the child both (*a*) assigned as many different number words as there were items and (*b*) tagged every item in the array. Why the second criterion? Some children skipped some items in the array but then double-counted a comparable number of other items; thus they assigned the right number of tags without tagging every item. Since we used videotapes, and since children usually pointed to an item as they tagged it, we could identify such errors. One might argue that whenever a child used as many distinct tags as there were items, even if he did so by making two different kinds of errors that canceled each other, he should be scored as correct. To so argue is to assume that the child recognized his own errors and deliberately made compensating errors. We prefer not to make such an assumption. In the analysis of Figure 8.1 we did allow such trials to be scored as correct. Here we want to determine the extent to which children honor the one-one principle in a precise fashion.

We obtained a high level of reliability on the assignment of children to each of the categories. After an assistant scored all the 3-, 4-, and 5-year-olds on this principle, another independent rater scored a random 20 percent of the children. The two scores agreed on 98 percent of the assignments for each set size.

The results of our analysis are shown in Table 8.5. The numbers in the *N* column indicate how many children could be scored for this analysis. A total of 21 3-year-olds, 19 4-year-olds, and 15 5-year-olds were tested. For set size 2 we scored 19 3-year-olds, 9 4-year-olds and

TABLE 8.5. Use of the one-one principle in the videotape experiment.

			Category of one-one score (percentage of subjects)			
Set size	Age group	*N*	Perfect[a]	Shaky[b]	Questionable[c]	No[d]
2	3-year-olds	19	73.6	15.8	5.3	5.3
	4-year-olds	9	100.0	0.0	0.0	0.0
	5-year-olds	5	100.0	0.0	0.0	0.0
3	3-year-olds	21	52.3	28.6	19.0	0.0
	4-year-olds	15	93.3	6.7	0.0	0.0
	5-year-olds	9	88.9	11.1	0.0	0.0
4	3-year-olds	20	35.0	30.0	20.0	15.0
	4-year-olds	18	77.8	16.7	0.0	5.5
	5-year-olds	15	66.7	13.3	20.0	0.0
5	3-year-olds	20	30.0	35.0	25.0	10.0
	4-year-olds	19	68.4	15.8	10.5	5.3
	5-year-olds	14	50.0	28.6	21.4	0.0
7	3-year-olds	17	23.5	0.0	47.0	29.4
	4-year-olds	17	58.8	17.6	17.6	5.9
	5-year-olds	15	46.7	33.3	13.3	6.7
9	3-year-olds	12	0.0	16.6	8.3	75.0
	4-year-olds	15	33.3	0.0	53.3	13.3
	5-year-olds	15	26.6	26.6	33.3	13.3
11	3-year-olds	11	27.3	9.1	9.1	54.5
	4-year-olds	17	17.6	5.9	64.7	11.7
	5-year-olds	15	40.0	20.0	33.3	6.6
19	3-year-olds	9	11.1	0.0	11.1	77.8
	4-year-olds	17	35.3	17.6	11.7	35.3
	5-year-olds	14	35.7	14.3	35.7	14.3

N indicates the number of children scored on a set size. Children who failed to count on a given set size (usually the small numbers), or who were unwilling to be tested on larger set sizes, or who inadvertently were not tested on a given set size could not be scored here. Thus the changing *N*'s. The largest possible *N*'s were 21, 19, and 15 for the 3-, 4-, and 5-year-olds, respectively.

a. Correct on at least 90 percent of trials.
b. Correct on 60–89 percent of trials.
c. Correct on 1–59 percent of trials.
d. Never correct.

5 5-year-olds. The rest of the children gave cardinal-only responses. By now we are familiar with this result: Only the youngest children are inclined to count such small set sizes out loud. Consider by contrast set size 19, for which we scored 9 of the 3-year-olds, 17 of the 4-year-olds, and 14 of the 5-year-olds. For this set size the problem

was not that the youngest children were giving cardinal-only responses. Rather, fewer 3-year-olds could be tested on sets of size 19 because of their tendency to withdraw from the experiment. Looking at the 3-year-olds' results on set size 19 makes it clear that even when we managed to test them they did not do well. A full 77.8 percent failed all their trials.

The results of this analysis are easily summarized. First, for each age group, fewer and fewer children tend to be scored as perfect as set size increases. The percentage of 3-year-olds scored as perfect drops off more sharply than do the percentages of older children so scored. Second, if we assume that children scored as either perfect or shaky on a given set size are able to apply the one-one principle for that set size, these results agree quite well with those from the magic experiments: The large majority of all children honor the one-one principle for set sizes of 2 to 5. Third, set size 7 differentiates the 3-year-olds from the older groups. For set sizes of 7 or larger, 3-year-olds are seldom perfect. In contrast, 4-year-olds do quite well on set size 7; if we combine the perfect and the shaky scores, 76.4 percent of them can be said to apply the one-one principle in this range. Finally, no fewer than 50 percent of the 5-year-olds can be credited with the one-one principle for set sizes of 9 to 19—a fact that leads us to conclude that by the age of 5 children are well on their way to generalizing the one-one principle to a large range of numerosities.

A puzzle appears when we look closely at the 4-year-olds' scores for the larger set sizes. Combining perfect and shaky scores yields the result that 33.3 percent, 23.5 percent, and 52.9 percent of the 4-year-olds use the one-one principle for sets of size 9, 11, and 19 respectively. We wonder why they do so much better on set size 19 than on set sizes 9 and 11. A careful inspection of the test sequences of these children provides one clue. For set sizes 9 and 11 we tested the children several times per set size. In the case of set size 19, we often restricted testing to one or two trials. Gelman (1972a) has reported that a young child's confidence in his ability to negotiate difficult tasks can be shaky. Repeated testing on such tasks can have the effect of leading the child to err. Thus we may have inadvertently undermined the children's confidence in their ability to count set sizes of 9 and 11—at least more than we did with set size 19. Some support for this conjecture comes from the fact that few 4-year-olds are scored as never correct on set sizes 9 and 11. Instead they tend to be scored as questionable, meaning that they were correct on at least one trial with each of these set sizes.

A rather striking effect appears across all ages for set sizes of 3 or larger. Children who were repeatedly asked to count a given set size began to stumble with respect to the one-one principle. It is possible that we forced them to falter by testing them over and over on the same set sizes. Alternatively, perhaps the children had not yet fully mastered the one-one principle for the various set sizes. In either case, the fact that repeated trials sometimes led to errors is clear. We think that this fact provides insight into the question of why the young child fails to conserve—even on small set sizes (Gelman, 1972b). If on different trials a child assigns different numbers of tags to the same set size, and if, as we assume (see Chapter 10), the young child uses a counting algorithm to reason about number, then he is bound to fail to conserve. A different number of tags has to lead him to report a different number of items.

Those familiar with the results of the magic experiment might well ask how we can offer the above account and yet find that young children treat small numbers as invariant in the magic paradigm. The magic paradigm was designed to minimize the likelihood of a child changing his estimate across trials. Further, children were never asked to count in Phase I. Obviously, they were not asked to count over again and thereby led to make mistakes. In any case, we have always viewed the results of the magic paradigm as evidence of whether children have *any* ability to reason about number rather than as evidence of whether they have perfect ability. We now begin to see why their competence is less than full. We will return to this issue later, for we have further reason to focus on the roles of counting and the one-one principle in the development of number conservation. But this discussion must await the presentation of more data.

Errors in applying the one-one principle. Even the oldest children we tested occasionally applied the one-one principle incorrectly. We were interested not just in whether they erred but also in what types of errors they made. As our method of transcription indicates, we were prepared for the possibility that children might make many different kinds of errors. This turned out not to be the case. Some tagging errors did occur, but they were few and were limited to one kind. The majority of the errors we observed can be classified as partitioning or coordination errors.

We expected to classify an error on a count trial as a tagging error if the child either repeated a tag, as in "one, two, three, two," or used an inappropriate tag. As already indicated, tagging errors seldom occurred. When they did, they always involved repeating a tag. We never observed the use of inappropriate tags (such as blue or a mouse). This outcome surprises us. After all, we did use arrays composed of heterogeneous items. Our youngest children might have been distracted by the heterogeneity and turned to labeling the objects by name or color. The fact that they did not further supports the conclusion we drew from the findings of Gelman and Tucker (1975), namely, that children as young as 3 can and do ignore the stimulus properties of an array when abstracting its numerosity. The following protocol shows a child who was interested in the colors of the objects but who managed to stay with the task of counting them.

C. D., age 3 years, 5 months, was being tested with four objects that varied in color.

How many are on that plate? *Um . . . Look, this is red and so is this. There's one, two, three, four.*

The rate at which tag-duplication errors occurred was low. In a total of 524 count trials, the 3-year-olds made only 12 tag-duplication errors. Half of these errors involved set sizes 11 or 19; the others were distributed across set sizes 2 through 7. The 4-year-olds were run on 525 count trials and made only 10 tag-duplication errors, all of them on set sizes greater than 9. Finally, the 5-year-olds made only 7 tag-duplication errors in their 398 count trials. As Table 8.6 shows, such errors were infrequent both absolutely and relative to other kinds of errors.

From these results, we conclude that children almost always assign distinct tags to items they are counting. They seem to know the component of the one-one principle that requires distinct tags. If they know this, why then *do* they err on tests of the one-one principle? Because, as Table 8.6 makes obvious, they have difficulty in partitioning items that are to be counted from items that are already counted. And because they have difficulty in coordinating the starting and stopping of the two processes necessary to the application of the one-one principle, partitioning and tag assignment. That is, they make partitioning errors and coordination errors.

We scored four types of errors as partitioning errors. The first type involved double-counting an item (or items) in the middle of an array. (Double counts of the first or last item were scored as coordination errors.) The second type occurred if a child returned to recount an item that he had already counted. These two types of partitioning errors are differentiated by when the child double-counted an item. In the first case the child assigned two tags to the same item before

TABLE 8.6. Rate (occurrence per 100 trials) of one-one errors in the videotape experiment.

				Type of one-one error		
Set size	Age group	Number of count trials	Number of error trials per 100 trials	Tag-duplication	Partitioning	Coordination
2–5	3-year-olds	402	32.6	1.0	21.6	21.9
	4-year-olds	256	10.9	0.0	5.8	5.1
	5-year-olds	191	18.8	1.6	11.0	6.3
7–19	3-year-olds	122	72.1	5.7	73.0	37.7
	4-year-olds	269	42.4	3.7	30.1	22.6
	5-year-olds	207	35.7	1.4	23.2	19.8

An error trial could have more than one error; hence the rates for the different kinds of errors do not sum to the rate of error trials.

going on to another one; in the second case the child assigned tags to items *x, y, z,* and so on, then went back to assign another tag to item *x*. The third type of error, the omission of one or more items in the middle of an array (not the first or last item) was classified as an omit error. Finally, the child who failed to count two or more of the final items in the array made the fourth type of partitioning error. We distinguish between the third and fourth types by where they occurred. Children who omitted items in the middle of an array gave the impression that they had lost track of which items they had counted. In contrast, children who left out several items at the end seemed not to realize that they were supposed to count all the items. The two kinds of omission errors occurred at different rates (see Table 8.7), with the latter being rare.

Table 8.7 allows us to characterize the partitioning errors. They occur most often when the child makes a slip in going from one item to an adjacent item; thus the high rates of double-count and omit errors. Children seldom stop before finishing a count or return to an item after subsequent items have been tagged. These tendencies lead us to conclude that children understand the demands of the partitioning process but are sloppy in meeting them.

In our treatment of the protocols from various magic studies, we classified errors of counting one too many or one too few items as coordination errors, because they seemed to reflect difficulties with starting and stopping. Inspection of the videotapes confirmed this hypothesis. A child would hesitate with his finger poised over the array, then abruptly start counting but point to the second or third item rather than the first. Another child would drum two or three times on the first item before moving off into an orderly partitioning sequence.

TABLE 8.7. Rate (occurrence per 100 trials) of partitioning errors in the videotape experiment.

		Type of partitioning error			
Set size	Age group	Double-count	Recount	Omit	Stop too soon
2–5	3-year-olds	9.7	2.8	8.9	0.2
	4-year-olds	3.1	0.8	1.9	0.0
	5-year-olds	2.1	0.5	8.4	0.0
7–19	3-year-olds	22.6	7.3	38.0	5.1
	4-year-olds	13.8	0.3	14.2	1.8
	5-year-olds	7.3	1.5	12.1	2.3

Similar sorts of lack of coordination occurred even more frequently at the end of the arrays. In scoring coordination errors we employed four criteria. First, a child could miss or double-count at the beginning of a sequence. Second, he could miss or double-count at the end. Third, he could commit what we call an overrun error, either by continuing to say number words after he stopped pointing or by continuing around a nonlinear array after counting each item once. Overrun errors typically occurred with nonlinear arrays; they probably indicated that the child did not realize he had counted all the items. Finally, the child could let his pointing get out of step with his tagging or vice versa, in which case we classified his error as asynchrony.

It can be seen in Table 8.8 that the majority of coordination errors came about because the child missed or double-counted the last item. Overrun errors occurred, but very infrequently. Asynchrony errors appeared with any frequency only when 3-year-olds were faced with larger set sizes. The 3-year-olds also had some difficulty with starting.

In summary, a child who fails to honor the one-one principle in the course of counting usually does so because he slips up in moving from item to item, thereby missing an item or tagging it twice. The two partitioning categories and two coordination categories that involve this kind of error account for most of the errors across all ages and all set sizes.

We take the results of these error analyses as further evidence that preschool-aged children's counting behavior is indeed guided by the one-one principle. We have already reported that the children in this study knew enough to use almost as many tags as they had items to count. We see from the error analyses that the tags assigned are al-

TABLE 8.8. Rate (occurrence per 100 trials) of coordination errors in the videotape experiment.

		Type of coordination error			
Set size	Age group	Beginning	End	Overrun	Asynchrony
2–5	3-year-olds	5.7	12.5	2.4	1.3
	4-year-olds	1.2	2.7	0.8	0.4
	5-year-olds	1.6	4.7	0.0	0.0
7–19	3-year-olds	5.8	21.4	1.6	8.8
	4-year-olds	0.0	19.0	2.6	0.9
	5-year-olds	0.0	14.0	4.8	1.0

most always distinct. Children seldom repeat the same tag, and they seldom stop before tagging all items in an array, although they do sometimes miss the last item. In short, their failures to honor the one-one principle indicate a lack of skill rather than the lack of an underlying conception or rule.

Evidence for the stable-order principle. Children whose behavior seems to follow the stable-order principle use the same list of numerlogs in the same order in different trials. We judged a child to be honoring this principle if (*a*) he used the conventional list of number words in the conventional order; or (*b*) he used an idiosyncratic list of numerlogs but used the same list repeatedly. If a child used the same list—either conventional or idiosyncratic—on 90 percent of the trials at a given set size, his performance was classified as perfect. (For any number of trials fewer than 10, this criterion required absolute consistency.) A child was classified as tending to honor this principle if he used the same list on at least 60 percent of the trials (but less than 90 percent) or if he sometimes used the conventional list and sometimes used an idiosyncratic list. In the latter case he had to use the same idiosyncratic list whenever he did not use the conventional one. All other children were judged not to have honored the stable-order principle.

The results of this analysis are simply summarized: Children of all three ages, if they are willing to try to count a given set size, will follow or tend to follow the stable-order principle. Only two children (both 3-year-olds) showed no clear tendency to apply this principle. Indeed, more than 90 percent of the 4- and 5-year-olds and 80 percent of the 3-year-olds used the same list on all of their trials, regardless of set size. Of course, not all 3-year-olds were willing to try to enumerate a set of 19 items; only 9 of our 21 3-year-olds made the attempt. This effect of set size on willingness to count was hardly noticeable in the 4- and 5-year-olds. Again we see that young children's application of the stable-order principle is not limited to small numbers. They occasionally err in applying this principle, but at no set size that they count do they completely fail to honor it.

In our analysis of the count trials in the magic experiment, we noted a tendency for children to use idiosyncratic lists of numerlogs. A comparable result holds in the videotape study. One 5-year-old used an idiosyncratic list on all set sizes. One 4-year-old used a conventional list on set sizes 2 and 3 but switched to an idiosyncratic list for set sizes 4–19. Another 4-year-old child introduced an idiosyncratic list just for set size 19. And, five of the 3-year-olds used idiosyncratic lists for the majority of their trials. Thus, some children, but not very many, do construct their own lists. We wish there had been more such children, because we had hoped to determine the relationship between set size and the tendency to use idiosyncratic lists. For now, we can simply note that when idiosyncratic lists appear, it is with younger children or larger set sizes.

Evidence for the cardinal principle. To what extent do children show an understanding that the last tag in a list of numerlogs has a special status, namely, that it represents the numerosity of the set? Our experiment was not optimally designed for answering this question. The procedure emphasized enumerating items rather than obtaining a correct numerical representation of the set of items. Schaeffer, Eggleston, and Scott (1974) used a procedure that is better suited to a precise evaluation of the child's use of the cardinal principle in counting: After the child enumerated the items, they covered the array and asked the child how many items it contained. The drawbacks of our procedure necessitate the following rather lengthy discussion of our scoring criteria.

The first two criteria for scoring a child as using the cardinal principle on a given set size require little comment. A child was said to use this principle if on one or more of the counts of a given set size he repeated the last numerlog in his enumeration (as in "one, two, three; three"), or if he enumerated the set on one trial and on a subsequent trial gave only the final numerlog from his previous enumeration. This second criterion precludes categorizing children as perfect or shaky on the cardinal principle depending on whether they use it on 90 percent or only 60 percent of their trials. Thus, a child was simply scored as using or not using the principle on a given set size.

Our third scoring criterion was the use of unusual stress on the last numerlog. As indicated earlier, we thought this use of unusual stress might be the earliest indication that a child was employing the cardinal principle after counting a given set size. We therefore expected this criterion to be met more often with younger children and with larger set sizes. This was indeed the case. Six 3-year-olds, two 4-year-olds, and one 5-year-old used exaggerated stress on the final numerlog in one or more of their counting sequences. And in 16 of the 23 counting sequences in which exaggerated stress appeared, the number of items being counted was five or more. However, as these numbers indicate, the use of exaggerated stress was not very common at any age on any set size (only 23 instances in several hundred trials). This fact casts some doubt on the hypothesis that exaggerated stress is the first indication of the use of the cardinal principle. The sparse occurrence of this behavior on the scoring sheets also, no doubt, reflects the fact that scorers were instructed to note it only if the stress was very loud and clearly unusual.

Our fourth and fifth criteria for crediting a child with the use of the cardinal principle are interrelated. Indeed, the fifth turns out to obviate the fourth. Since some readers may have reservations about the fifth criterion, however, we did the analysis with and without it. The fourth criterion, which is relevant only in the absence of the fifth, is the use of the cardinal numerlog alone. Despite the fact that they were asked to count, many of the 5-year-olds simply looked at the smaller sets (2–4) and, without overt counting, said, for example, "There's three." Of course, scoring such behavior as evidence for the use of the cardinal principle in counting reifies the hypothesis that the numerlogs are obtained by covert counting rather than by direct apprehension (subitizing). Not to admit this criterion, however, would produce distinctly odd results. We would have to conclude that 5-year-olds are more apt to use the cardinal principle on a set of 9 than on a set of 2. Worse yet, we would have to conclude that on set sizes of 2 and 3, 3-year-olds are more likely to use the cardinal principle than 5-year-olds. The fact that only older children show cardinal-only behavior encourages the assumption that this behavior reflects covert counting, not subitizing. This assumption is further encouraged by the fact that every single child who gave cardinal-only responses to small sets was scored, by other criteria, as using the cardinal principle when he encountered larger sets.

This last point brings us to our fifth criterion. It strikes us as odd to assert that a child uses the cardinal principle in counting a seven-item set but not in counting a three-item set. Yet according to the scoring sheets, several children, particularly younger ones, appeared to do just that. Thus, we decided to credit a child with the cardinal principle on a given set size if he showed by other criteria that he used it on a larger set size. The inclusion of this criterion (called the *inferred* criterion in Table 8.9) also provides us with a device for scoring the few children who were not tested on a particular set size. This criterion makes the previous one unnecessary, because every child who gave cardinal-only responses when asked to count smaller sets met one of our first three criteria when counting larger sets. Thus, if the use of the cardinal principle on sets of size 5, 7, 11, and so on is taken to indicate the capacity to use it on smaller set sizes as well, then all of the children who were scored as using the cardinal principle on the smaller sets would have been scored that way even without the fourth criterion.

Most of the children who were scored as using the cardinal principle on a given set size were so scored on the basis of the two most straightforward criteria: They repeated the final numerlog at the termination of a count, or they counted on one trial and stated the final numerlog on a subsequent trial. These criteria, plus the third one, exaggerated stress, were the only ones that came into play on the larger sets; and the third criterion, as already noted, seldom figured in the scoring.

The percentages of children of different ages who used the cardinal principle are plotted as a function of set size in Figure 8.3. Recognizing that some readers may not be persuaded by our justifications for the third, fourth, and fifth scoring criteria, we show in Table 8.9

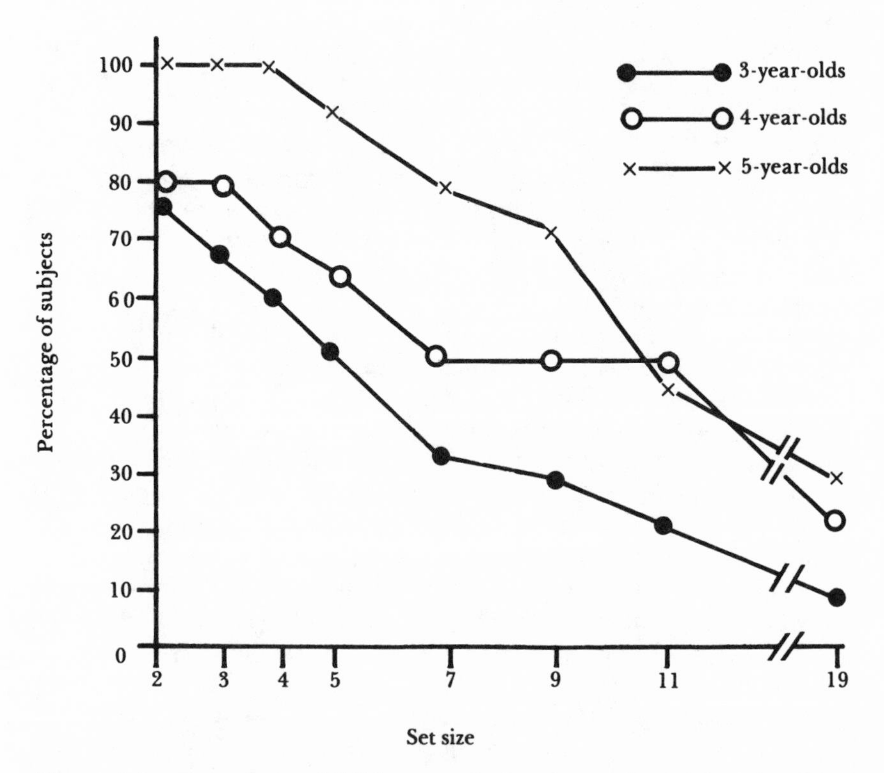

Figure 8.3. Percentage of children employing the cardinal principle at the conclusion of an enumeration, as a function of set size.

TABLE 8.9. Use of the cardinal principle in the videotape experiment.

		Percentage of subjects who met cardinal criteria				
Set size	Age group	All five	All save stress	All save cardinal-only	All save inferred	N^a
2	3-year-olds	76	71	67	67	21, 21
	4-year-olds	79	79	32	78	19, 18
	5-year-olds	100	100	33	100	15, 14
3	3-year-olds	67	62	67	52	21, 21
	4-year-olds	79	79	58	79	19, 19
	5-year-olds	100	100	47	100	14, 15
4	3-year-olds	62	57	62	40	21, 20
	4-year-olds	68	63	68	67	19, 18
	5-year-olds	100	100	100	100	14, 15
5	3-year-olds	52	29	52	40	21, 20
	4-year-olds	63	63	63	63	19, 19
	5-year-olds	93	93	87	93	14, 15
7	3-year-olds	35	25	35	24	20, 17
	4-year-olds	50	44	50	53	18, 17
	5-year-olds	80	80	80	80	14, 15
9	3-year-olds	31	13	31	33	16, 12
	4-year-olds	50	50	50	47	18, 15
	5-year-olds	73	73	73	73	14, 15
11	3-year-olds	23	15	23	18	13, 11
	4-year-olds	50	44	50	41	18, 17
	5-year-olds	47	47	47	47	14, 15
19	3-year-olds	11	11	11	11	9, 9
	4-year-olds	24	24	24	24	17, 17
	5-year-olds	29	29	29	29	14, 15

a. The percentages in the first three columns are based on the first *N*. It represents actual observations plus the inferred observations. The second *N*, representing only the actual observations, was used to compute the percentages in the fourth column.

the effect of dropping each of these. The results agree with those obtained from the magic experiment. Each successive age group has more children who honored the principle after longer counts. At least 60 percent of the 3-year-olds did so when counting set sizes 2–4; 60 percent of 4-year-olds did so with set sizes 2–5; and 60 percent of 5-year-olds did so with set sizes 2–11. For set sizes of 7 or greater, most 3-year-olds seldom honored the principle. The older children honored it reasonably often for all the set sizes except 19.

A comparison of Tables 8.5 and 8.9 suggests that the cardinal principle is harder for the children to follow than the one-one principle. We had a hint to this effect in the magic experiment results. To be better able to discern whether this is the case, we need to consider the extent to which children coordinate their use of all three how-to-count principles.

Composite profile analyses. So far we have considered children's counting behavior principle by principle. We find considerable evidence that they are able to follow the one-one and stable-order principles, even on the largest set sizes. Children frequently err in applying these principles, but our analysis of the kinds of errors suggests that they err not because they fail to understand requirements of the principle but because they lack the skill to meet those requirements. This suggestion leads to the question of how children treat the remaining principles when the performance demands of one or more principles exceed their capabilities. To answer this question and the question of the relative difficulty the child encounters in attempting to honor each principle, we analyzed the extent to which the principles were used in conjunction with one another. In making this analysis we ignored our distinction between perfect and shaky use of the one-one and stable-order principles. A child was scored as honoring the one-one principle in counting a given set size if he did so on at least 60 percent of the trials. Similarly, he was scored as honoring the stable-order principle if he used the same list of numerlogs in the same order on at least 60 percent of the trials. The results of this analysis are presented in Table 8.10. Note that some of the entries in the column labeled "All three" are followed by second entries in parentheses. The initial figure represents the percentage of children who counted aloud *and* used all three principles. The second figure shows the percentage of children who used all three principles, under the assumption that children who gave the correct cardinal answer without counting aloud applied all three principles.

This analysis makes clear both the nature and the course of the development of counting in children. This development, at least from the age of 3 on, involves the perfection of skills rather than the apprehension of new principles. When the set size is small and the demands on the child's skill are therefore minimal, most 3-year-olds apply all three principles in conjunction. That is to say, they count correctly. As set sizes become larger and the demands on skill correspondingly greater, the children typically stop applying the cardinal principle. That is, they stop giving evidence that they regard the last numerlog as a representation of the numerosity of the set. Thus, as Table 8.10 indicates, when children use only two of the principles rather than all three, the two they use are the one-one and the stable-order principles.

That children stop using the cardinal principle first suggests that they have some ability to monitor their own skill at counting. The last numerlog represents the set's numerosity only if one has in fact adhered to the one-one principle and the stable-order principle in arriving at that last numerlog. By the time the child stops using the cardinal principle, he is having difficulty following the other principles, particularly the one-one principle. We suggest that the child stops using the cardinal principle when he becomes uncertain about his performance on the other two principles.

Table 8.10 provides evidence that the one-one principle causes more difficulty than the stable-order principle as set size increases. As the table shows, when children adhere to only one of the three principles, it is generally the stable-order principle. It might be thought that children who use only the stable-order principle are merely rattling off numerlogs without attempting an actual enumeration. This is not the case. A review of the videotapes to check this possibility made it clear that, with extremely rare exceptions, all of the children who used only the stable-order principle on larger set sizes were trying to enumerate the sets. One indication of this is the fact, already noted, that on 98 percent of all counting trials the children pointed from item to item as they counted. When children fail to honor the one-one principle in counting large sets, it is not for lack of trying.

Additional evidence that difficulty with the one-one principle is related to the child's reluctance to use the cardinal principle comes from the more differentiated analysis of which Table 8.10 is a distillation. Remember that in Table 8.10 children are said to use the one-one principle for a given set size if they did so on at least 60 percent of their trials on that set size. In the more differentiated analysis, we distinguished between children who were perfect (at least 90 percent ad-

TABLE 8.10. Combinations of counting principles in the videotape experiment.

Set size	Age group	Combination used (percentage of subjects)							Not run
		All three	Two out of three			One out of three			
		One-one, stable-order, and cardinal	One-one and stable-order	One-one and cardinal	Stable-order and cardinal	Cardinal	Stable-order	One-one	
2	3-year-olds	67 (76)[a]	19	—	—	9	5	—	—
	4-year-olds	27 (74)	21	—	—	47	—	—	5
	5-year-olds	33 (93)	—	—	—	60	—	—	7
3	3-year-olds	67	24	—	—	—	9	—	—
	4-year-olds	58 (79)	21	—	—	21	—	—	—
	5-year-olds	60 (100)	—	—	—	40	—	—	—
4	3-year-olds	57	9	—	—	—	29	—	5
	4-year-olds	68	21	—	—	—	5	—	5
	5-year-olds	100	—	—	—	—	—	—	—

5	3-year-olds	43	24	—	9	—	19	—	5
	4-year-olds	63	21	—	—	—	16	—	—
	5-year-olds	86 (93)	7	—	—	7	—	—	—
7	3-year-olds	19	9	—	14	—	38	—	19
	4-year-olds	47	21	—	—	—	21	—	11
	5-year-olds	80	7	—	—	—	13	—	—
9	3-year-olds	—	9	—	14	—	33	—	43
	4-year-olds	37	11	—	5	—	26	—	21
	5-year-olds	67	7	—	7	—	20	—	—
11	3-year-olds	5	19	—	5	—	24	—	48
	4-year-olds	37	5	—	5	—	42	—	11
	5-year-olds	47	27	—	—	—	27	—	—
19	3-year-olds	—	5	—	5	—	28	—	62[b]
	4-year-olds	16	37	—	5	—	32	—	11
	5-year-olds	20	33	—	7	—	33	—	7

a. Number in parentheses reflects percentage of children who can be said to have applied all three principles if we assume that cardinal-only children should be so characterized.

b. Includes one child who was run and judged to have failed to apply any principles.

herence) or shaky (less than 90 percent but at least 60 percent). The tendency for children not to use the cardinal principle was clearly accompanied by an increase in the percentage who were shaky in their adherence to the one-one principle.

Table 8.10 shows once again that the tendency to simply state the cardinal number without counting aloud is confined to the very small numbers and grows stronger between the ages of 3 and 5. There is no doubt that the 4- and 5-year-olds who show this cardinal-only behavior have mastered the application of all three counting principles to these smaller sets. Indeed, cardinal-only behavior appears to emerge as a consequence of that mastery. Whether this implies that cardinal-only behavior is a consequence of covert counting or of some newly emerging subitizing skill is hard to discern. We return to this question in a later chapter. In any event, we believe that the older children who show cardinal-only behavior with the smallest sets should be regarded as competent to apply all three counting principles in enumerating such sets. The effect of including them in the group of those who honor all three principles is shown by the percentages in parentheses in Table 8.10.

In summary, an analysis of what combinations of the three counting principles children adhere to at each set size yields the following conclusions: (1) At small set sizes (2–3) children adhere to all three principles; (2) as set size increases they begin to have trouble with the one-one principle, and they stop using the cardinal principle; and (3) in enumerating the largest sets they try to adhere to the one-one principle but fail, while continuing to adhere with fair success to the stable-order principle. This pattern is true regardless of age, although, not surprisingly, the younger the child the smaller the set sizes at which he begins to falter. A child-by-child analysis of performance as set size increased confirmed these conclusions, which were originally drawn from group percentages.

This pattern of findings leads us to the view that at a very early age children know the fundamentals of enumeration. They know that enumeration requires the use of a stably ordered list of tags. This knowledge appears to motivate their learning to recite the conventional list of numerlogs. It explains the occasional appearance of stably but idiosyncratically ordered lists of conventional numerlogs. It also explains the occasional use of stably ordered lists composed of tags not ordinarily used for enumeration, most notably the alphabet. Children further understand very early that enumeration involves placing the list of numerlogs in one-to-one correspondence with the items that are to be enumerated. They understand this necessity for coordinating the stable-order principle and the one-one principle, but they are not very good at putting this understanding into practice, particularly when confronted with large sets. These conclusions are buttressed by the data on 2-year-olds, which reveal the presence of both stable ordering of lists and components of the one-one principle.

Component Counting Skills in 2-Year-Olds

As noted earlier, our method of videotape data reduction did not work for the tapes of 2-year-olds. Recall that we required coders to make up to 30 entries per count trial. When we attempted to code the data on 2-year-olds we had the impression that we were forcing decisions about how to code a trial. Rather than do this we thought it best to transcribe the videotape sessions fully and to await the outcome of the analyses of the older children's trials. The hope was that studying the older children would provide clues about what to look for in the 2-year-olds' transcripts. Now that we know more about the skills that are involved in the application of the counting principles, we can ask whether some of these skills are manifested in 2-year-olds' efforts to meet our task requirements.

We had a sample of 19 2-year-olds available for testing. Two of these children were dropped from the sample because they seemed to have no idea of what we wanted them to do. Another child's data were lost through equipment failure. This left us with 16 children who ranged in age from 23 to 35½ months (median age 2 years, 5 months). These children were run on as much of the planned procedure as possible in the same fashion described for the older children. No 2-year-olds were tested on set size 19, and only 7 were tested on set size 11. The numbers of 2-year-olds who were tested on the remaining set sizes were 16, 15, 13, 12, 11, and 7 for set sizes of 2, 3, 4, 5, 7, and 9, respectively.

Do 2-year-olds have any appreciation of what is involved in counting? Apparently they do. All but one of the children used lists of number words. And the evidence that they followed the stable-order principle is striking. Considering the lists that each child used over repeated trials on a given set size and across set sizes, we can identify three classes of children: One class used idiosyncratic lists; another used conventional lists; and one child used the same number word over and over again (as in "three, three, three"). The 39 percent of the children who used idiosyncratic lists had conventional lists for the first three entries and then took up with their own lists (for example, "one, two, three, six, five, ten"). One of the children who used idiosyncratic lists recycled when he ran out of number words, creating a list that went "one, two, three, one two, three," and so on. The most interesting fact about the children who used idiosyncratic lists is that they adhered perfectly to their chosen order over trials. Of the nine children who tried to use the conventional order of numerlogs, only one did so perfectly. The others fell into the category of shaky on the stable-order principle, that is, they used the same list on at least 60 percent of their trials. No child was observed to be simply spouting number words in random sequence, although one child seemed to have available only the number word *three.* Note that children who adopted idiosyncratic lists used these consistently, while all but one of the children who adopted the conventionally ordered list had some difficulty in using it. Apparently children can follow the stable-order principle more easily with an order they have created than with the conventional order. Why? Again, the explanation we offer is that the young child has a principle in search of a list. The child who follows the stable-order principle in his own fashion might be characterized as the child who allows his own convictions to win out over the dictates of convention. He seems to have good reason to do so; for he succeeds in meeting the demands of the stable-order principle.

What do 2-year-olds do while reciting these stably ordered lists? Do they have any inkling of the fact that the numerlogs are to be assigned as tags to items in a set? In other words, do they have any understanding of the one-one principle? We think so. One finding that supports this conclusion is that all children pointed on at least some trials. Not too surprisingly, their tendency to point was somewhat sporadic; only 6 of the 16 children systematically pointed across trials. Still, 10 children pointed on enough trials for us to be able to determine the kinds of one-one errors they made. Of these 10 children, 9 made partitioning errors and 7 made coordination errors. (Some children made both types of errors and some only one type.) Recall that coordination errors involve double-counting an item at the beginning or end of an array or omitting the last item in an array. The fact that such errors occurred led us to consider the possibility that the children were assigning approximately as many tags as there were items to tag.

A second line of evidence also supports our view that 2-year-olds have some idea of the one-one principle. As should be apparent by now, 2-year-olds who were asked to identify or count the number of items in a given array responded by reciting number words. They did not necessarily use conventional lists of words, and one child seemed content with a string of threes. But did they come close to reciting as many number words (distinct or not) as there were items in the arrays? To answer this question we proceeded as follows: For each child we counted the numerlogs recited on each trial with a given set size. We then computed a mean for each set size for each child. We chose to use the mean in order to represent extreme tendencies in the summary statistics. In a sense, to use the mean was to work against ourselves—a strategy that we prefer to adopt whenever we are seeking evidence of abilities that are generally presumed not to exist. Such a strategy should make any order we find in our results more convincing. As it turns out, in this case the results do show some order.

Figure 8.4 shows the range of the means for set sizes two through nine, the set sizes for which we were able to obtain a considerable amount of data. The dot within each range shows the mean of the

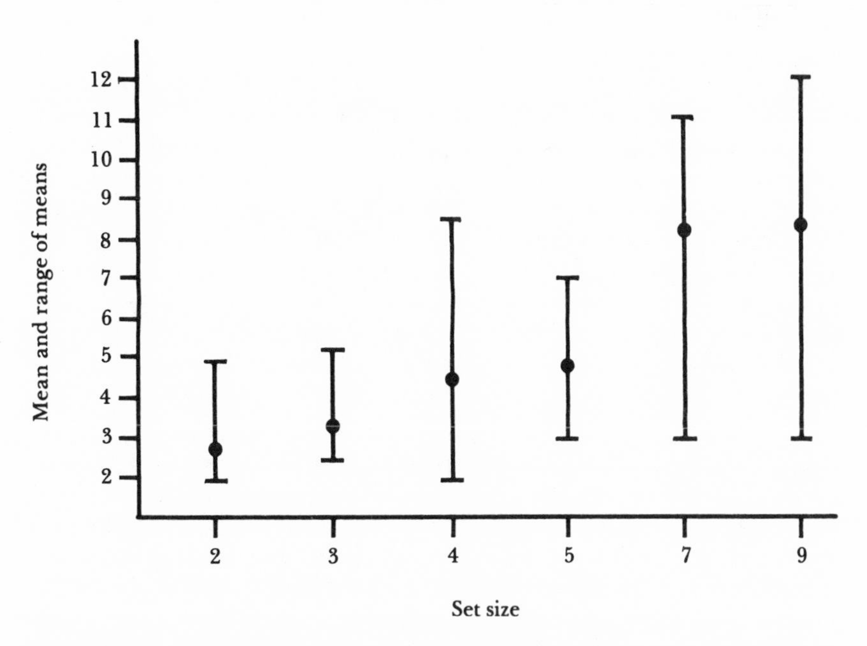

Figure 8.4. The range of the mean number of tags used by individual 2-year-olds in counting a given set size, shown as a function of set size. A vertical bar indicates the range of the means; the dot within the range indicates the group mean (the mean of the individual means).

means—the central tendency figure. Several features of the figure are noteworthy. First, there was an overall tendency for the children to use more tags as set size increased. The correlation is imprecise, however. The children distinguished between set sizes two and three. They further distinguish these set sizes from set sizes four and five; however, the central tendency figures for set sizes four and five are almost identical (4.5 and 4.8 respectively), suggesting that the children had difficulty discriminating between these two sizes. Recall that these arrays were shown at the same time. Likewise, the central tendency figures for set sizes seven and nine are virtually identical (8.2 and 8.3 respectively.) Still, on the average more tags were assigned to these larger arrays than to arrays of four and five items. As compared to older children, the 2-year-olds were not very good at assigning a number of tags that matched the number of items in an array. Nevertheless, it can hardly be said that they assigned tags randomly. The larger the numbers represented by the pair of displays in front of them, the more numerlogs they recited. How could this be true unless the children at least had some notion that to count means to assign number words as tags to items in an array?

In sum, we find, as others have (for example, Schaeffer, Eggleston, and Scott, 1974), that 2-year-olds have already begun to master the sequence of count words. Unlike Schaeffer and his colleagues, however, we do not think that this acquisition of number words is isolated from attempts to enumerate. The evidence shows that the young child attempts to honor the one-one principle. To be sure, young children are not very successful in this attempt, but they do try. How could they even try if they did not have some understanding of the one-one principle?

We also analyzed the 2-year-olds' data for evidence of ability to follow the cardinal principle. Eight (50 percent) of the children were able to identify the cardinal numerosity of a two-item array. All of these children could count the same number when asked to do so, and three of them held up two fingers when giving the answer "two." Only four children identified the cardinal numerosity of three-item arrays; these children could also count this set size.

Like the older children, the 2-year-olds showed a weaker tendency to follow the cardinal principle on set size two in the videotape study than in the magic experiments. Of the 16 children scored in the [(2 vs. 3) − 1] magic protocols, 14 gave some evidence of using this principle. Once again we suspect that the explanation is related to the fact that in the videotape trials we repeatedly asked the children to count.

In addition, we note that the videotape sample includes some very young subjects. The median age of the children who did *not* provide evidence of being able to identify the cardinal numerosity of set size two was 27 months; the median age of those who did provide such evidence was 31 months. All the children who followed the cardinal principle on set size three were at least 32 months old; indeed, three of them were 35 months old. Still older children fared considerably better on set sizes two and three. It seems that the younger the child, the less likely he is to give any evidence of following the cardinal principle. This particular result does not surprise us. It is, however, noteworthy in another context. In Chapter 6 we reviewed theories claiming that children subitize small sets before they can count. According to such theories, our 2-year-old subjects—especially the very youngest of them—should have been able to correctly answer the question "how many?" when shown an array of two items and possibly even an array of three items. The role of a subitizing mechanism in making numerical judgments will come up again in Chapter 12. For now, we simply note that all of our videotape results show a positive correlation between age, counting proficiency, and the ability to state the cardinal numerosity of an array—even arrays of only two or three items.

What have we learned about 2-year-olds? To be sure, they are not skilled counters. Nevertheless, they do use some components of the counting procedure. They attempt to tag items, they point (albeit in a less-than-systematic fashion), and they appear to be guided by the stable-order principle. They have much to learn. But they will not be without help, for they seem to have available counting principles, principles that guide them in their efforts to achieve performance mastery. Perhaps now we understand why young children seem to be so caught up in counting rituals and so very interested in numbers.

Chapter Nine

The Abstraction and Order-Irrelevance Counting Principles

We have seen that preschoolers' counting behavior is guided by principles that tell them how to count. As children get older, they learn to apply these principles to larger set sizes. The question yet to be answered is whether preschoolers also use the abstraction, or "*what-to-count,*" principle and the order-irrelevance, or "doesn't matter," principle.

The Abstraction Principle

The how-to-count principles are procedural in that they specify the way to execute a count. The abstraction principle has little to do with procedure; instead, it deals with the definition of what is countable. It delineates the domain of objects or events that can be counted. As noted in Chapter 7, adults seem to assume that any object or any event in any form or representation can be collected together with any other and then counted. Whether the items to be counted belong to the same semantic category, whether they are the same color, whether they are all two-dimensional or three-dimensional, whether they exist or are imagined matters not one bit. A count can even be made of sets of a given count; for example, two piles of five objects can be counted as two fives. Indeed, it seems that almost any combination of things can be counted.[1] This fact is so obvious to adults that they are likely to be surprised that it needs to be mentioned at all, let alone elevated to the status of a counting principle. Why do we see the need to treat it specially? The answer comes from a consideration of some theoretical accounts of counting—accounts that imply that

1. It is not quite true that any imaginable set of entities can be counted. For example, the points that make up a line cannot be counted.

young children do not recognize that all discrete items, real or imagined, are candidates for counting.

Gast (1957) advanced one theory of the development of counting skills that highlights the possibility that young children do not know the obvious. He suggests that initially the child can only count items that are completely homogeneous and three-dimensional. This amounts to saying that compared to the adult the young child places severe restrictions on the definition of what constitutes a countable collection. Klahr and Wallace (1973) likewise hold that the definition of what is countable is initially very restricted. They propose that children first learn to apply the counting procedure to objects that have common perceptual properties, such as color and size. Only later do children realize that objects of the same identity but different color and size can be counted in the same collection. It is still later that they apply the counting procedure to very heterogeneous sets. Gast also suggests that young children place restrictions on the mode in which an object is represented: thus the hypothesis that the ability to count two-dimensional materials develops later than the ability to count three-dimensional materials. Gast does not mention the ability to count imaginary things or even things that cannot be presented visually (such as great minds of the nineteenth century). We can only conjecture that if he had concerned himself with such representations he would have assumed that the ability to count them would develop even later than the ability to count objects represented in two dimensions.

Developmental accounts like Gast's are tied theoretically to a particular view of number concepts, the view that number concepts develop along with the child's ability to classify objects and events into organized hierarchies. It is a common idea that children first classify together objects that share salient perceptual properties and only later develop the ability to use "abstract" criteria for the purposes of classifying. The theme that concepts are initially perceptual and then abstract (conceptual, logical, and so on) dominates the major theoretical writings in developmental psychology (Bruner et al., 1966; Piaget, 1952; Werner, 1957). Piaget explicitly ties the development of complex classification skills to the development of a concept of number. Thus, traditional assumptions about the linkage between classification schemes and the concept of number have fostered the belief that children place restrictions on what can be counted. We do not rule out the possibility that this is the case. Indeed, these theories led us to think that it was important to obtain evidence about what children regard as countable. We entertained the hypothesis that the domain of events and objects to which young children would apply the how-to-count principles might be severely restricted. We found, however, that preschoolers do not ordinarily place restrictions upon countable collections.

It would be surprising if preschoolers placed *no* limits on what can be counted. We might expect, for example, that they would find a request to count a class of imagined events bizarre. Since no evidence about their reactions to such a request is available, we can do no more than suggest this possibility to make clear that we do not think it out of the question that some things adults consider countable might not be considered countable by young children. Still, we contend that the existing evidence rules out the claim that young children place severe restrictions on what constitutes a countable collection. Over and over again, we found that even our very youngest subjects were willing to classify together and count diverse items.

Later in this chapter we report on a counting experiment using trinkets that were three-dimensional representations of a wide variety of objects (whistles, chairs, rings, animals, abstract shapes, and others) and that varied in size and color. Not a single child referred to an object by its name or some perceptual attribute in response to a request to count. This result was what we expected. Recall the findings of the videotape counting experiment, in which we used objects that varied in color. Not one 3-, 4-, or 5-year-old child made the tagging error of designating an item by its color. All children assigned number words as tags. We did not report on a similar analysis for the 2-year-olds, in part because of the coding difficulties. The reader might conjecture that the 2-year-olds made such errors. In fact, they did not. All of the 2-year-olds assigned number words as tags. Only one of the 16 2-year-olds showed any inclination to interrupt his recitation of count words and turn to talking about the objects. On one of his trials this child counted an array of four items as follows: "One, two, three, a blue." In short, we recorded only one instance of a failure to use number words or letter words in counting diverse and frequently heterogeneous sets. It seems likely that if young children restricted their definition of countable collections to items that are perceptually alike they would hesitate to use number words as tags when faced with collections of dissimilar objects. We found no such tendency. We also did not observe children picking out items that were the same in some way (such as color) and counting just those items—another strategy that could be adopted by a child who restricted the domain of countables to items that resemble one another.

Even though heterogeneity in an array does not affect young children's tendency to assign number words as tags, it might affect the accuracy of their performance. Our results indicate that it does not. First, the how-to-count results from the magic protocols, which involved homogeneous materials, agree closely with those from the videotape experiments, which involved heterogeneous materials. Second, Gelman and Tucker (1975) investigated the extent to which heterogeneity in a two-dimensional array interfered with the ability of children of ages 3–5 to provide absolute judgments of numerosity and found that the children in the heterogeneous conditions did as well as those in the homogeneous conditions. Finally, in magic experiments involving surreptitious changes in the identity or color of an object, Gelman and Tucker found evidence that young children consider such changes irrelevant to the numerosity of a set. Thus, in several different experimental settings, young children have been shown to be as willing to count heterogeneous items as to count homogeneous items. And their accuracy seems to be unaffected by the introduction of diverse materials. Given the range of tasks and stimulus materials involved, we are confident in concluding that the young child can apply the counting procedure to sets of diverse composition.

What are we to make of the findings that seem inconsistent with this conviction? Gast did report that heterogeneity had an adverse effect on the young child's ability to accurately assign a numerical representation to a set. Klahr and Wallace (1973) cited Gast's findings in support of their model of the development of number concepts. Siegel (1973) has observed an adverse effect of heterogeneity in a discrimination-learning task that required children to respond to a difference in numerosity in order to receive a reward. We do not doubt these results; we do, however, question their interpretation.

The children in Gast's study were tested on homogeneous sets before being tested on heterogeneous sets. Gelman and Tucker assigned their subjects to either a homogeneous or a heterogeneous condition but not both, on the hunch that Gast's procedure might have set the children to think that only like items were to be enumerated together. In a follow-up study, Gelman and Tucker (1975) demonstrated that they could influence whether a child focused on the overall numerosity of a heterogeneous set or on the numerosity of like items within the set. Children who are tested first on homogeneous materials and then on heterogeneous materials will sort the heterogeneous materials into homogeneous subsets and count the items in each subset. If the opposite sequence of testing is followed, that is, if heterogeneous arrays are presented first and homogeneous arrays second, heterogeneity has no adverse effects on performance. It appears that young children like adults, are sensitive to context. When the context requires or seems to require that only like items be included in the enumeration, children restrict their application of the counting procedure to those items. We suggest that the studies that report developmental functions for stimulus materials and modes of representation should be thought of as studies of the child's ability to respond to context. To be sure, this ability will be related to their abilities to invoke different modes of representation, to classify, and the like. But it does not follow from such a line of argument that young children start with a bias toward restricting the domain of countables. Indeed the evidence speaks to the contrary. Young children can readily group a variety of two-dimensional and three-dimensional materials together under the rubric of "things to be counted."

Siegel's results can be reinterpreted for different reasons. She compared learning rates with homogeneous and heterogeneous materials. In the context of discrimination-learning tasks, heterogeneous displays involve more irrelevant dimensions than do homogeneous displays. It is a well-established fact that the introduction of irrelevant dimensions slows down the learning of a solution (see, for example, Trabasso and Bower, 1968; Gibson, 1969). It is also known that the younger the child, the more likely he is to be distracted by irrelevant characteristics (Gibson, 1969). It should come as no surprise then that Siegel reports an interaction between age and type of display: The younger her subjects, the more heterogeneity slowed down their learning rates. They did learn, however; it cannot be said that they were unable to abstract numerosity from a heterogeneous array. We believe that they took longer than the older children not because classifying unlike objects together was inherently more difficult for them but because their attention was diverted by other attributes of the heterogeneous sets.

In general, then, we hold that young children honor the abstraction principle and are able to treat a wide variety of objects as countable. We do not have much evidence about the range of applicability of the counting procedure. We do know that 3- and 4-year-olds will treat two collections of items as numerically equivalent even if they contain different kinds of items. Consider the behavior of children in magic experiments who are asked to fix the game after the experimenter has made a surreptitious subtraction. Often they ask for another item. We hand them more than the number requested to see if they pick just one. Inadvertently, we give them an opportunity not only to restore the expected number of items (say three) to the altered plate but also to turn the loser plate (which contains say, two items) into another winner plate. Many 3- and 4-year-old children take advantage of this opportunity. They add an item (or items) to both plates, then announce with glee "Two winners! Two threes!" As often as not, the two winners contain dissimilar items.

In concluding that young children honor the abstraction principle, we side with Macnamara (1974), who rejects the notion that the development of counting abilities and the development of classification abilities are necessarily yoked. To count heterogeneous materials, a child simply needs to treat them as *things.* He need not know how to assign these things to various levels of a classification hierarchy. Only on tasks that require the application of such knowledge does its absence become a problem. To say that failure on tasks that require complex classification skills occurs because a child believes that only like objects can be counted is to confound one cognitive ability with another. When care is taken to avoid this mistake, the evidence clearly favors the view that preschoolers can treat diverse items as things and proceed to count them. The main question that remains to be investigated is how wide a range of things (real or imagined) the young child will count

The Order-Irrelevance Principle

In a sense the order-irrelevance principle is redundant to the other counting principles. It captures the way the first four principles interact in contributing to a full appreciation of counting. To understand what counting is all about, an individual must understand that the order in which items are tagged doesn't matter. The following example illustrates what we mean.

Assume that we have an adult count an array of heterogeneous items. These items are toy-sized versions of a chair, a dog, a baby, a flower, and a car. They are arranged in a row from left to right in the order listed. Likely as not, the adult will start at the left or right end of the row, and count the first item as one, the next as two, and so on, until he reaches the last item in the row and calls it five. This done, we scramble the array and once again place the items in a row. This time the objects are arranged from left to right as follows: flower, car, baby, dog, chair. Asked to count again, the adult tags the flower as one, the car as two, and so on, and ends up tagging the chair as five. Next, leaving the objects in the same positions, we point to the car and say "Start counting with this as number one." Our adult subject tags the car as one, the baby as two, the dog as three, and the chair as four, then backs up to tag the flower as five. In later counts we ask that the car be the three, the four, and the five. The adult responds by skipping around the array to include all the items in each count.

What have we shown by this demonstration? First, we have shown that our hypothetical adult knows that each of the count words can be assigned as tags to any of the items in the array. Further, he knows that the order in which items are partitioned and tagged does not matter. What does matter is that the tags themselves be assigned in a fixed order and that each item be assigned a distinct tag. Which item receives a given tag is irrelevant. The tags do not attach themselves to a particular object; they are arbitrary ways of designating an object, which for purposes of a count is a *thing* (the abstraction principle). Therefore, a given tag can be reassigned to any other object in subsequent trials. Despite this reassignment of tags, the same count and cardinal number result no matter what order items are processed in, because (*a*) in any one count each item is tagged once and only once; and (*b*) the tags are always withdrawn in order from the same stably ordered list; with the result that (*c*) one always obtains the same final tag as the representative of the set's numerosity. As this explanation indicates, the recognition that it does not matter which tag is assigned to which item constitutes an explicit recognition of the abstraction principle and an implicit recognition of the other three counting principles.

If the "doesn't matter" principle were no more than a composite of the other four principles, we would be best advised to use the illustrated behavior as evidence for the ability to coordinate all the principles. Indeed, such behavior could not occur without the application of all four principles, but we think that it implies something more than the ability to coordinate the how-to-count and what-to-count principles. Namely, it demonstrates an understanding of the fact that much about counting is arbitrary. Thus, the choice of "doesn't matter" as an alternate label for this counting principle.

The order-irrelevance principle deals with knowledge about the consequences—actually the lack of consequences—of the particular way the first four principles are applied. It says that the representation of the numerosity of a set that one obtains by counting is invariant with regard to the order in which the items in the set are counted. The individual who behaves as if "it doesn't matter how you count them as long as you do count them all and count each only once" shows an awareness of this invariance.

PILOT STUDIES

We asked young children to count arrays in much the same way that our hypothetical adult subject did. The reader may be surprised at our decision to do so. For the request to count by making a certain object the number one then the number two, and so on must confront a young child with a formidable task. First, he may not understand the instruction itself. It is quite vague and leaves much room for misinterpretation. Second, if the child understands what we want him to do, he has to skip around the array as he assigns tags, perhaps to think of moving the items in order to assign a particular tag to a particular object, and so forth. Before we instituted a standardized procedure we asked ourselves whether we had any evidence that might support a decision to do so. Then, we ran pilot studies.

As we have pointed out many times, the magic experiments elicited extensive counting behavior. The children frequently counted the same set several times in the course of an experiment. In doing so they seemed to pay little or no attention to the order in which they counted the objects. This lack of attention to order was particularly noticeable in experiments involving heterogeneous arrays. A given child making repeated counts started the count with one item on one occasion and another item on another occasion. Some children even seemed to play at counting and recounting the set, with the apparent goal of starting each count with a different object. This apparent lack of concern about the order in which the items of an array were enumerated suggested to us that the child was following the order-irrelevance principle. Such observations influenced the design of subsequent counting experiments.

In the videotape experiment children were asked repeatedly to count the same set of objects. This was done partly to allow us to determine the extent to which children followed the stable-order and one-one principles. We also wanted to know whether children would assign a given tag to more than one item. Thus the decision to rotate and scramble the arrays. With videotaped data we could see whether a child tried to maintain his original tag assignment. Would the child who initially tagged a given item as *one* continue to do so trial after trial, or would he reassign the tags as the objects were rearranged?

The children seemed to be completely indifferent to which tag was assigned to which item. Rotating the plates resulted in no discernible tendency for children who had counted from left to right to switch and count from right to left. More importantly, each time the arrays were scrambled, the children took a different object as the starting point for their count; that is, on each trial the children reassigned tags. It is of particular interest that these results held even for our youngest subjects. Recall that these youngest children used number words as tags. Despite the fact that they made a variety of errors in applying the how-to-count principles, they did *not* use the names or attributes of objects as tags. We conclude that children "know" that in counting the relation between the tag and the object tagged is arbitrary.

The reader might object that we make too much of this apparent indifference about which object is tagged with a given numerlog. Yes, the behavior is congruent with an assumption that the children honor the order-irrelevance principle. They count the same objects in a different order, thereby reassigning tags. But do they understand the many principles that together justify this indifference to order? The behavior under discussion does not allow us to conclude this. Such a concern led Gelman to design pilot procedures that instructed children to systematically reassign tags.

Gelman ran two pilot studies. The first involved 19 3-year-olds and 25 4-year-olds in a two-step procedure. In the first step the experimenter had her subjects count sets of from four to seven heterogeneous objects over and over again. The items in the sets were rearranged between counts. This part of the experiment showed us how well a child applied the how-to-count principles. It also allowed us to check once again whether young children are indifferent to which numerlog is assigned to which item. The second pilot study was run as an additional phase of the videotape counting experiment described in Chapter 8. Those children who were willing to stay on were tested with the same kinds of questions as those used in the first pilot study. For these subjects we had no need to obtain an assessment of ability to count or of indifference to the order in which objects were tagged. The major videotape experiment had served these purposes.

Both pilot studies continued with questions designed to probe the child's explicit understanding of the "doesn't matter" principle. This phase of the studies was run as follows: A child was shown a linear array of four or five trinkets that varied by item type, color, and size. To start, the experimenter asked the child to count this set. Almost invaribly, the children responded by counting the row from left to right or vice versa. The experimenter then asked the child to start the count with an item in the middle of the row. The instruction to the child was "make this number one". The experimenter went on to ask the child to make a particular object the two, the three, or even the four. Several other questions were tried to get a feel for what questions we could reasonably include in the full-scale study: "Does it matter how you count?" "This time you said this is one, and just before you said *that* was one. Is it okay to have two things that are called number one?" "Can any of these be one (two, and so on)?" "Can you start counting anywhere?" For each question the experimenter also attempted to get the child to explain or justify his answer.

As already indicated, the children in the videotape experiment seemed indifferent to their order of tag assignment. The same is true of the children in the first pilot study of the order-irrelevance principle. All but 1 of 44 children freely assigned different tags to the same object as it was moved about during the initial count trials. The one exception was a young 3-year-old (39 months), who tried to tag a given object with the same numerlog as the display was rearranged. Perhaps this unwillingness to change an object's tag is the reason for this child's failure to honor in a consistent fashion any of the how-to-count principles.

So far we have reported nothing new. We have simply confirmed the young child's basic indifference to the issue of which item receives which tag. We still do not know whether this indifference reflects an explicit and articulable understanding that tag assignments in a count are arbitrary, temporary assignments. The following protocols indicate that it clearly did so for some of our subjects. As before, the children's words appear in italics.

K. G., age 4 years, 10 months.

Can you count and make this [an object in the middle of a nonlinear array] number one? *Yes.* [K. G. demonstrates.] Could this [another object] be number one? *Yes. All of them could.* Could it be number two? *Yes. All of them could be—six, four, and five and any number.* Why? [K. G. demonstrates by moving objects into different positions for a series of counts and showing that the number names are being reassigned as she does this.]

J. W., age 4 years, 6 months.

Can you make this [an object in the middle of a linear array] number one? [J. W. moves the designated object to the beginning of the row and then counts with it as one.] Why not start in the middle? *No, I don't like to start that way.*

[Experimenter asks J. W. to make another object one. J. W. moves the object to the beginning of the array. This happens once again. Then the experimenter scrambles the array and asks J. W. to make an object in the middle number one. Leaving the array as is, J. W. does so and skips back and forth until all objects are counted. The experimenter then asks him to make the same object the five. J. W. succeeds again and proceeds to demonstrate that he can make all of the objects number five.]

Is it okay to call two different things five? *Yes.* Why? *You change them around.* [He points in different ways to illustrate. Experimenter then places 10 objects on the display.] Can you make all of them ten? *Sure, watch.* [He demonstrates.] Can any be ten? *Any you wanted.* Can any be number five? *Of course!*

S. V., age 4 years, 0 months.

Make this [an object in the middle of a five-item array] number one. [S. V. starts with that object and counts those to the right but does not return to the two objects on the extreme left.] *One, two, three.* Can you make this [another object] number one? [This time S. V. performs perfectly, counting the item in fourth position as one, that in fifth position as two, and then those in first, second, and third positions as three, four, and five.] Does it matter how you count them? *No.* Why? [No answer, but subject counts the same array twice, assigning the tags differently on each trial.] How come there are two number ones? Now you say this is one, and before that was one. [No answer.]

A. S., age 4 years, 7 months.

[A. S. had followed the experimenter's instructions to make a particular object in a three-item display the one, two, and three.] Is it okay to make this different numbers? *Yes because you took that one off and . . .* How can you call this number one if this is number one? *Easy!* Look, first you said this was one, then this, then this. That means that there's just one, because this is a one, and this is a one, and this is a one. Right? *Oh, come on!*

A. M., age 3 years, 9 months.

Make this [the third in a row of four items] number one. *One, two* [the third and fourth items], *one, one* [the first and second items], *no . . . that's not right.* Okay, we'll do it again with this [the second item] as one. *One, two, three* [second, third, and fourth items], *four* [first item]. Now make this [the third item] one. *One, two* [third and fourth items], *three, four* [first and second items]. Can two of these be called one? *No, this* [first item] *is number one.* Is it always number one? *No.* Could any be number one? *All of them!* Can they all be any number? *Yes!*

Of course some children did not do as well as those just quoted. Some needed a considerable amount of prompting; others had trouble figuring out what the experimenter wanted of them; and still others seemed to have no idea what to do. These various levels of difficulty are illustrated in the following protocols:

R. M., age 3 years, 8 months.

Make this [an item in the middle of a five-item linear array] number one. [R. M. starts with the designated object] *One, two . . . wh, I can't count any more.* Make this [another object in the middle of the array] the one. Start counting here. [She starts with the designated object] *One, two* [last two items], *three, four, five* [first three items in the array]. Does it matter how you count? *No.* Why? [No answer.]

N. G., age 3 years, 1 month.

Make this [in a linear array] one. [N. G. points to the object] *One.* What about the rest? [N. G. does nothing.] Count them, make this number one. *One, two* [last two items in the row]. What about the rest? [With extensive help and probing he finally succeeds.] Is it okay to call two things one? *Yeah, these two are number ones.* Can anything be called two? *Yes, okay.* Why? [No answer.] Three? *Yeah.* Why? [No answer.] Can they be any number? *Yeah.* Why? [No answer.]

N. G., age 3 years, 7 months.

Make this number one. *One.* Make this [another object] number one. *One, two* [counts from designated object to end of the row and leaves out the remaining two items]. Make this [the same object as on previous trial] number one. [This time the child succeeds.]

D. S., age 3 years, 4 months.

Make this number one. [D. S. touches the object.] Can you start counting here and make it number one? *One, one* [two successive objects] *what's that?* [Experimenter repeats the question. This time D. S. starts with the designated object and counts to the end, leaving out the remaining objects.] Can you make this be two? *Two.* Does it matter how you count? *Yes.*

It is no accident that the protocols that show children having difficulty with the "doesn't matter" questions involved 3-year-olds. In general, the older the child, the better was his performance. The majority of 4-year-olds gave clear evidence of understanding the order-irrelevance principle. Although some 3-year-olds did the same, most did not. Still, we draw attention to the fact that most 3-year-olds attempted to answer the questions. This led us to keep children from this age group in the experiment proper.

One further result from the pilot studies deserves mention. All of the children were given grades for their ability to apply the how-to-count principles in concert. Comparing each child's ability to count with his ability to do well on the "doesn't matter" tasks yielded two results of interest. First, all the children who showed explicit understanding of the order-irrelevance principle were reasonably good counters. By reasonably good counters we mean children who were scored as having used all three how-to-count principles. We do not mean children who never made counting errors. Children who were shaky (see definition in Chapter 8) in their use of the one-one or stable-order principle but nevertheless used all three principles could do very well on the order-irrelevance principle. Although all the children who did well on the order-irrelevance tasks were good counters, not all those who were good counters did well on the order-irrelevance tasks. In other words, it appears that being a reasonably good counter is a necessary but not a sufficient condition for getting a high score on the "doesn't matter" test.

The Order-Irrelevance Experiment

Having discovered through the pilot studies that it would be possible to assess a child's level of understanding of the order-irrelevance principle, we turned our attention to obtaining a systematic data base. We were interested, of course, in whether the age trends and relationships that had appeared in the pilot studies would prove reliable.

The order-irrelevance experiment included the kinds of questions used in the pilot studies. First, however, we wanted to determine how willing our subjects were to reassign tags. In particular we wanted to discover whether the children knew that a given object could be assigned any of the count tags as long as the other objects were tagged in a systematic fashion. Accordingly, each child was asked to count a set several times, making a particular object first the *one,* then the *two,* then the *three*—and so on up to the set size.[2] After a child had been asked all tag-reassignment questions pertaining to the designated object, he was asked some of the same questions about a different object in the array. When children completed the second set of questions they were confronted with the fact that they had tagged two different objects with the same number words. We reasoned that being so confronted might help children make clear the fact that they knew that tag assignments are arbitrary. Telling a child that he had used the same tag or tags for two different objects allowed us to introduce yet another set of questions that might evoke explicit verbalizations about the order-irrelevance principle.

Readers who are familiar with research on the development of language objectivity may expect us to find that children do *not* realize that number words are temporary designates of objects rather than names for particular objects. When children of preschool age are asked if it is possible to change the names of the sun and the moon, or a dog and a cat, they say no (Piaget, 1929; Osherson and Markman, 1974–75). This unwillingness to change the names of the objects in their environment is taken as evidence of a lack of language objectivity. The children seem not to know that the names are arbitrary labels of objects and that calling a round shining spot in the daylight sky the *sun* is merely a convention. How can we expect young children to recognize the arbitrary status of number words if they do not recognize the arbitrariness of names of objects?

Upon reflection, we discern a difference between number words and names of objects. By definition, that is, by virtue of the counting process, number words must be arbitrary. It must be possible to use them as tags for counting any collection of objects (the abstraction principle). On the other hand, although our assignment of the names for objects is a matter of convention, a child can reasonably expect a particular object to be called by the same name over and over again. A failure to recognize that naming conventions are arbitrary does not stand in the way of being able to communicate about objects. Thus, the number names and other types of names are arbitrary for differ-

2. After a child was asked to make the designated object be N, he was then asked if he could make it be $N + 1$ (one more than the actual set size). These data are not presented here.

ent reasons. The arbitrariness of number names derives from the definition of their use in the counting procedure. The arbitrariness of our names for objects is a fact about language and society.

We decided to take advantage of this difference between number names and object names. Recall that we planned to ask children to count so that a given object (say a baby doll) would be tagged in turn as *one, two,* and so on up to the size of the set. Likewise we planned to have children count the same array so that a given object (say a toy chair) would be tagged in turn as *one* and *two.* After a child did this we could point out that he had, at different times, called both the doll and the chair the same thing, say *one* or *two.* Then we could ask if it were possible to switch the names of the chair and the baby and call the chair a baby or vice versa. We expected the children to say that the names could not be switched, in which case we planned to ask why it was all right to change an object's number word but not its name. We had no way of predicting how young children would deal with such a challenge, but we considered it worth presenting. We thought it might elicit explanations about the reassignment of number tags.

In sum, the order-irrelevance experiment was conducted for several reasons. First, we wanted to confirm our pilot findings regarding the order-irrelevance principle, namely that it constitutes part of the counting knowledge of good counters. Second, we sought a clear demonstration of the ability to systematically reassign a pool of tags to one object. Third, we thought it advisable to demonstrate the ability to assign the same numerical tags to different objects. Finally, we wanted to make a concerted effort to elicit explanations.

The subjects in this experiment were 48 preschoolers, 16 from each of three age groups. The median ages of the 3-, 4-, and 5-year-olds were 43 months, 52 months, and 66 months. Some children were run through the tasks in their homes, others at a nursery school. In all cases, the experimenter spent time playing with each child before running the experiment proper.

The experiment consisted of four parts. In Part I the child was asked to count a set of five objects on six different trials. Each trial involved a different arrangement of the same objects. The child watched the experimenter rearrange the array. The arrays were linear in half the trials and haphazard in the other half. When the child had completed the six trials, the experimenter covered the array and asked how many objects it contained. If a child had difficulty counting five objects, Part I was repeated with four objects. This part of the experiment provided the necessary data for assessing the extent to which a given child applied the three how-to-count principles. Note the adoption of Schaeffer, Eggleston, and Scott's test (1974) for the availability of the cardinal rule at the end of the counting trials. Children who seemed to be able to count five objects were tested with a five-item array in the next part of the experiment. Other children were tested with a four-item array. Requests for counting were made by one of our puppets, Mr. Lion or Mr. Worm.

Part II of the experiment was designed to test a child's willingness to (*a*) assign different tags to the same object and (*b*) assign the same tags to two different objects. It started with the experimenter placing four or five objects in a row so that from left to right a toy baby was in the second position and a toy chair was in the next-to-last position. The experimenter asked the child if he wanted to show Mr. Lion some tricks. Not too surprisingly, all children said yes. The experimenter proceeded by saying, "First, you have to count these." After the count the experimenter said, "Now here's the trick. Start counting with the baby and make it number one." If the child hesitated the experimenter said, "Count them all but make this [pointing to the baby] number one." This trial completed, the experimenter next asked the child to make the same object be number two, number three, and so on. An attempt was made to test a child on all of these trials, no matter how he did, stopping only when he refused to continue. At various times between trials, the child was told that he was doing very well—whether or not he was.

Having completed testing on the baby, the experimenter pointed to the chair and said, "Now let's do this trick another way. Make the chair be number one." In this case, the experimenter tried to get the child to designate the chair by at least two different tags, so that the experiment could proceed.

The third and fourth parts of the experiment were designed to elicit explanations about the arbitrariness of tag assignments. In Part III, the experimenter drew attention to the fact that a child had assigned the tags one, two, three, and so on first to one object and then to another object. The child was asked if this was okay. If he said yes, he was asked if it would also be all right to exchange the names of the baby and the chair. The child's response determined what the experimenter did next. Questioning continued until a child either verbalized some distinction between number-tag reassignment and object-name reassignment or seemed unwilling to continue.

The fourth part of the experiment began with the experimenter saying that Mr. Lion (or Mr. Worm) wanted his turn to show tricks.

Then the puppet voice was used to ask the child to count a row of four or five chips of different colors. The puppet then moved the first item to the end of the row and asked the child to count again. Once more the item in first position was moved to the end of the row and the child was asked to count. This process continued until all of the chips had been in the first position and counted as number one. Then the puppet piped in with "I tricked you, I made you make them all number one!" The session ended with the experimenter turning to the child and asking if this was so and how it could be.

The proceedings of the entire experiment were videotaped for later transcription.

Counting ability. We marked the children's skill at counting by giving grades. Children who showed the coordinated use of all three principles and never erred were given a grade of *A*. Those who made counting errors but used all three principles and were at least 60 percent correct on the one-one and stable-order principles were given a *B*. Children who coordinated their use of the one-one and stable-order principles for a set of five items but only gave evidence of applying the cardinal principle on smaller sets received a *C*. Those who applied the one-one and stable-order principles but not the cardinal principle received a *D*. A grade of *E* was assigned to children who honored only the stable-order principle or none of the principles.

As Table 9.1 shows, the older children were better counters. All of the 5-year-olds, 87 percent of the 4-year-olds, and only 56 percent of the 3-year-olds were scored as good counters (grades *A, B,* and *C*). These results are comparable to those reported in Chapter 8.

As before, we noted little systematic tendency to assign the same tag to the same objects over the six count trials with a given array.

Tag reassignment ability. The scoring of a child's performance on

TABLE 9.1. Counting ability in the order-irrelevance experiment (percentage of subjects).

	Age group		
Grade	3-year-olds	4-year-olds	5-year-olds
A	6	62	81
B	37	19	12
C	12	6	6
D	31	12	0
E	12	0	0

Part II of the experiment, the part that required a child to make an object other than the first one in a row be number one, number two, and so on, was done in two steps: First we rated success on each of the separate trials; then we rated overall success across all trials.

Performance on each separate trial was assigned to one of five levels, representing different amounts of ability to follow our requests to tag a designated object with a particular tag and count all objects. A child who simply counted from left to right, ignoring the request to tag the specified item with the specified numerlog, or who assigned the specified numerlog to the specified object but made no attempt to count any of the remaining objects was scored at level 5 for that particular trial. Children who tagged the designated object as requested but neglected to count the items to the right or left of the designated object were scored at level 4 for that trial. Children who tagged the designated object as requested but made counting errors of the kind described in Chapter 8 were scored at level 3 for that trial. Children who, in order to meet our request, adopted strategies that forced them to make counting errors were assigned a level 2 score for that trial. Finally, children who both met our request and counted without error were given a level 1 score for that trial. The following protocol illustrates a level 2 performance.

J. S., age 4 years, 3 months, is being tested on a five-item linear array. The child is asked to make the baby, the fourth item in the array, be the five. He begins by labeling the first item *one,* skipping the second item, and labeling the third, fourth, and fifth items *two, four, three—no, that's not it!* He starts again, holding his finger over the row while saying *one, two, three, four,* then touching the baby as he says *five.* Next he tags the third, second, and first items *six, seven, eight,* then returns to the fifth item and ends his count with *nine.*

J. S. tagged all items with unique tags, did not double-count or omit objects, and used the standard count sequence. The only trouble is that he used nine rather than five tags. He did so because he started counting with the baby. To start with the baby *and* meet the request that the baby be tagged as five, he adopted the strategy of reeling off the first four numerlogs without using them as tags. J. S.'s protocol contains the germ of another level 2 strategy that we observed. Note that on his first attempt J. S. reversed the standard numerlog series to say "one, two, four, three." Some children adopted this inversion strategy when asked, for example, to make the second object in an array be number three. They would count the array by saying "one, three, two, four, five." These children ended up with the correct final numerlog by deliberately committing two violations of the stable order, violations that cancel each other. This is a very sophisticated strategy, which from other perspectives might not be treated as an error at all.

Children who received level 1 scores on a particular trial typically did one of two things. Some left the objects in position and skipped around as they counted. Others picked up the designated object and moved it to the position it would have to be in to receive the specified tag when the child counted the array linearly from one end to the other. The following portion of S. S.'s protocol illustrates the latter level 1 strategy.

S. S., age 4 years, 3 months.

> The child is asked to make the baby (the fourth of five items) be the five. She starts at the first item and tags the first three items *one, two, three.* Then she says *I'll have to put the baby there* while she lifts it out of the fourth position and moves it to the left of the fifth item. Then she continues to count *four, five.*

As is clear from the preceding protocols, many of our subjects adopted complex problem-solving strategies. We will not elaborate on this fact here. For a detailed account of such strategic behavior, see Merkin and Gelman (1978).

Having reliably scored the performance on each trial at level 1, 2, 3, 4, or 5 (agreement between two independent raters was almost perfect), we assigned each child an overall performance rank of I, II, III, IV, or V. A child's overall performance rank reflected the number of level 1 scores on his individual trials. Level 1 scores on all trials led to an overall rank of I; level 1 scores on at least 60 percent (but not 100 percent) of the trials led to a rank of II; level 1 scores on two trials led to a rank of III. Children with but one level 1 score received an overall rank of IV, and children without any level 1 scores received a rank of V. Those whose rank was V had considerable trouble with this part of the experiment; children who earned no level 1 scores earned no level 2 scores either. The only exception was one child who received three level 5 scores and two level 2 scores.

As Table 9.2 shows, age had a pronounced effect on overall performance rank. A full 94 percent of the 5-year-olds, 69 percent of the 4-year-olds, and only 31 percent of the 3-year-olds received overall

TABLE 9.2. Overall performance ranks for tag reassignment in the order-irrelevance experiment (percentage of subjects).

Overall performance rank	Age group		
	3-year-olds	4-year-olds	5-year-olds
I	12	44	75
II	19	25	19
III	12	6	6
IV	12	25	0
V	44	0	0

ranks of I or II. Almost half (44 percent) of the 3-year-olds earned the lowest rank; none of the 4- or 5-year-olds did this poorly.

One might conclude that the 3-year-olds who received rank V on the order-irrelevance items had no understanding of what the experiment was about. This is not the case. Of the seven children in this age group who earned rank V, two seemed not to understand the task. The other five all received at least one level 2 or level 3 score on their individual trials. That is, they succeeded, at least once, in tagging the designated object as requested and attempting to count all other objects. The problem is that they made counting errors. This fact raises the question of what relationship exists between the child's counting ability, which was assessed in Part I of the experiment, and his overall performance on the order-irrelevance tasks. The data are given in Table 9.3.

The table makes it quite clear that children who scored as better counters tended to receive the top two overall performance ranks. Some good counters earned low overall ranks, however. Thus a child's ability to perform the order-irrelevance tasks, reassigning tags without making sloppy counting errors, is indeed related to his ability to count. But the ability to apply the how-to-count principles in con-

TABLE 9.3. Relationship between counting ability and overall performance rank on order-irrelevance task (number of subjects).

Counting ability	Overall performance rank				
	I	II	III	IV	V
Better counters (grade *A*, *B*, or *C*)	19	9	2	4	5
Poorer counters (grade *D* or *E*)	1	0	1	1	4

cert is not sufficient to insure mastery of the order-irrelevance principle.

Explanations. One goal of the order-irrelevance experiment was to obtain explanations from our subjects. We gave children two opportunities to explain the order-irrelevance principle. In the first they were asked whether two different objects could be tagged with the same numerlog and whether the common names of two different objects could be interchanged. If they behaved as if the number names could be reassigned but declared that the objects' names could not be, they were asked why the two cases were different. In the second situation they were confronted with the fact that they had been "tricked" into assigning every object in the array the same numerlog in different count trials, and they were asked how that could have happened.

An inspection of the protocols obtained during the first questioning session made it clear that our subjects recognized some difference between reassigning numerlogs and reassigning labels. This recognition is well illustrated in the following excerpts from protocols.

R. H., age 3 years, 11 months.

You said you can call this number one and this number one. Does that mean you could call this [the baby] a chair? [screeching] *No-O-o!* Why not? *I want to put the baby in the chair.* Do you think you can call this [the baby] an airplane? [R. H. shakes head no.]

A. C., age 3 years, 0 months.

This [the baby] could be number one Right? *Okay.* And this [the dog] could be number one? *Yeah.* Is it silly to call this [the baby] a doggie? *Yeah, cause it's not a dog.*

M. L., age 3 years, 10 months.

Remember when you made the airplane number one and the baby number one? [M. L. nods yes.] Well, if this can be number one or this can be number one, does that mean that this [the baby] can be called an airplane? *No, because it's a baby.*

We obtained similar evidence from 10 (of 16) 3-year-olds, all 16 of the 4-year-olds, and 15 of the 5-year-olds. Thus, as expected, our subjects said that numerlogs could be reassigned but resisted the suggestion that the names could likewise be reassigned. Nevertheless, we failed to elicit an explanation of the order-irrelevance principle. No child gave a clear account of how numerlog reassignment is different from name reassignment. The children who answered questions about this difference offered explanations of why the names could not be reassigned; they did not seem to realize that we wanted an account of why numerlogs *could* be reassigned. We suggest that the source of trouble was our question format rather than a complete lack of ability on the part of the children to explain the order-irrelevance principle. More than half of the children did provide explanations during the second questioning session.

Eight of the 3-year-olds, 10 of the 4-year-olds, and 14 of the 5-year-olds responded to the challenge to explain how each object in turn could have been called number one. The typical explanation addressed the fact that the objects had been rearranged. This being so, a different object could be assigned to the first position in the array and therefore called number one. The following protocols illustrate the ability of preschoolers to explain their ready reassignment of tags over trials.

D. P., age 3 years, 11 months.

This time the blue one [chip] was number one, but before that the red one, and before that the white one. How could they all be number one? *I don't know.* Why do you think? *Because their name can't all be number one.* They can't? *Except when you're counting, they can.* How can they all be number one? Is this the way we count: One, one, one. *No! You give them each a number but not the same number. You can change when you move them.*

R. H., age 3 years, 11 months.

You called everything number one; how? Can they all be number one? *I can move them.* Could they all be number two? [R. H. nods yes.] You said this was number one, then you said this was number one; how can they all be number one? [R. H. moves the chips around three times to show that he calls the object in first position, no matter what it is, number one.]

N. F., age 5 years, 6 months.

How can they all be number one? *Because you changed them around like this* [N. F. moves the chips.] And what happens? *Well this* [last object in row] *was number one but you changed them around.* Can they all be number one at the same time? *No.* Why not? *Because one, two, three—that means at one time. They can't be all ones unless you count them one, one, one . . .* When you count them one, one, one is that all right? *No-O-o!*

D. K., age 3 years, 8 months.

How could they both be number one? [D. K. counts the chips again.] But you said the white was number one and before the blue; how could they both be number one? *No-o-o! But that's how you're supposed to do it because they're supposed to be number one each time, different times.*

G. S., age 4 years, 5 months.

Mr. Lion says he fooled you—you called everything number one. How can they all be number one? *Because if you move them around, you have to start with one.*

T. T., age 4 years, 10 months.

[After counting three times] *I know how many there are!* Before you said this was number one, then this, then this, how can they all be number one? *Because you keep on changing them.* What happens if you change them? *Well, then you'll have to move this* [the first one on the left]. *Look!* [While taking away some objects] *If you didn't need these, all there would be is two.* [He points to the two remaining objects.] *See now like this would be number one and this would be number two.*

This last child seemed to be telling us that the number name of an object was temporary—so temporary that if one removed the object designated *one,* some remaining object would have to be called *one.* Thus he showed that if one rearranged *or* removed objects then the assignment of tags would start from the beginning. The reader might object to our taking such explanations as evidence of the young child's ability to articulate the order-irrelevance principle. What else might the children have said? What might you, the reader, have said? When we ask adults similar questions they give much the same answers as our subjects.

We do not mean to claim that preschoolers understand the order-irrelevance principle as well as adults do. Even if adults failed to give an explanation, we would expect them to achieve an overall performance rank of I on the tag-reassignment task. Although many of our subjects did well on this task and provided explanations of the principle, many others did not. Most children had some notion of what is involved, but they clearly had further to go before they would reach a full understanding of the order-irrelevance principle. Still, they understand more than we thought they would when we timidly began our pilot studies. What is more, those who understand—those with overall performance ranks of I or II—also are able to tell us what it is that they understand. We *do* find more capacity than meets the eye when we take the trouble to look.

Chapter Ten

Reasoning about Number

We have discussed the principles of the counting process that the young child uses to abstract a property of arrays, namely their numerosity. Now we turn our attention to the principles that govern the way the young child thinks about this property once he has abstracted it. Our discussion is motivated by the results of Gelman's magic experiments, which were described at length in Chapter 8. In the magic experiments, Gelman built up expectations in her subjects about two arrays of different numerosity by a series of identification trials in the guise of a shell game. Then either the numerosity or some number-irrelevant facet of the "winning" array was surreptitiously altered. In attempting to reconcile what they actually found on the magic trial with what they expected to find, the children revealed a rich set of numerical reasoning principles.

In our description of the reasoning principles we assume that the counting process is definitional. That is, we assume that the system of reasoning principles takes the numerosities of sets as defined through the counting process. The numerical reasoning principles operate on representations of numerosity (Gelman, 1972a). These representations are either the cardinal numerons produced by applications of the counting process or perhaps, in some instances, the entire sequence of numerons produced by an application of the counting process. It follows that the numerical reasoning principles we outline below come into play only in situations where the counting process can be more or less relied upon to work.

We have already seen that very young children do not use the counting process reliably with set sizes greater than three or four. Thus, when set size becomes appreciably greater than four, we may expect the young child to become less and less disposed to apply his numerical reasoning principles, because the representations upon which those principles operate become less and less reliably defined.

It is important to emphasize that the counting process is not an intrinsic part of the reasoning principles. Rather, counting provides the representations of reality upon which the reasoning principles operate. That is, counting serves to connect a set of reasoning principles to reality. Not surprisingly, the child's behavior reflects both the reasoning principles and the counting procedure that mediates between them and reality.

All the evidence we deal with now comes from experiments with small set sizes, because only with small sets is the link between reasoning principles and reality reliable. With larger sets the representation a child gives for the numerosity of a given set changes from one enumeration to the next even when nothing about the set has changed. Our initial task is to describe the reasoning capacities the young child *has* in a domain in which they are clearly at work—limited though that domain may seem at first. Later we ask how these competences enable the young child to develop more extensive and complex reasoning processes.

The Numerical Reasoning Principles

The magic studies provide our basic source of data regarding the way young children reason about number. The child's responses to the unexpected changes in Phase II of the magic experiments involve integrating his representation of the Phase II events with his established representation of the Phase I events. For example, the child who in Phase II encounters a different number of items than expected will say he no longer wins because the number is less (or more) than it should be.

Recall the basic findings of the magic experiments (Bullock and Gelman, 1977; Gelman, 1972a, 1972b, 1977; Gelman and Tucker, 1975). Children as young as 3 (and sometimes $2^1/_2$) years behave as if they know that transformations involving displacements do not alter number. They also seem to know that transformations involving addition and subtraction do alter the numerical value of an array. Further, subjects seem to know that changing the color or identity of items in an array does not alter number; that the number pairs 1 and 2 and 3 and 4 can be ordered and can be compared so as to relate 3 to 1 (both exemplars of the *less* relation) and 4 to 2 (both exemplars of the *more* relation); and that addition and subtraction cancel each other. Finally, although young children know that subtraction and addition can undo each other, they do not necessarily know that the addition of *X* items is undone by the subtraction of exactly *X* items.

These are the facts. What must we assume to explain them? We organize our answer to this question around the kinds of reasoning principles that could guide the young child's ability to integrate numerical representations in relation to the potential or actual effects of transformations. We contend that three kinds of numerical reasoning principles are available to the young child: relations, operations, and principles of "reversibility"

The Relations

Equivalence. In reasoning about number, the young child recognizes an equivalence relation. The evidence behind this statement comes primarily from those magic experiments in which the winning array was transformed in a fashion that was irrelevant to number. In these experiments, children unexpectedly encountered arrays that had been lengthened or shortened or in which an item of different color or identity had been substituted for one of the original items. In nearly all cases, the children regarded the altered array as equivalent to the original array, that is, as still the winner. When the children who noticed the transformation were probed about the reason for their equivalence judgment, they characteristically indicated that, although other attributes of the array had changed, the number had remained the same. In other words, the equivalence of the altered array to the original array rested on the equivalence of their numerosities. For example, a child might say, "They moved out. It still wins. It's three now and it was three before."

A second line of evidence comes from the magic studies reported in Gelman and Tucker (1975). When asked to reverse an identity-change transformation, half the children constructed *two* dissimilar arrays with numerosities equivalent to that of the original winner plate, and said that they now had two winners. The following protocal illustrates this type of behavior.

D. S., age 4 years, 7 months, participated in an experiment that involved starting with a three-item heterogeneous array—two green mice and a soldier—and a two-item homogeneous array—two green mice. The three-item array was designated the winner. In Phase II it was altered to produce a three-item homogeneous array—three green mice.

[Experimenter asks subject how to fix the game.] *You take this* [a

> mouse] *off and put on a soldier. Where's a soldier?* [Experimenter gives subject extra objects, including soldiers.] *How about two winner ones?* [D. S. places the soldier on the two-mouse plate.] *This is gonna be a winner plate too. Both have three things.*

Notice that D. S. spontaneously constructed an equivalence based on numerosity rather than simply recognized one.

It is difficult to understand what would lead children to ask whether the numerosity of the altered set equaled the numerosity of the original set if the principles guiding their reasoning about numerosity did not include an equality relation. It is even more difficult to understand what would prompt the child to construct another set that was equivalent to the original set in number but in few other properties. Thus we assume that the child's behavior is guided by a principle that says that two numerosities do or do not satisfy a numerical equivalence relation.

This assumption leaves entirely open the procedure or algorithm by which the child decides whether two numerosities he encounters in the real world do or do not satisfy the equivalence relation. There are two reliable ways of establishing whether two finite sets are equivalent. One is to show that they both yield the same representation of numerosity. In this case the judgment of the equivalence in numerosity is mediated by representations of numerosity. The other procedure involves showing that a one-to-one correspondence can be established between the items of the two sets. This procedure is almost universally taken as definitional in formal developments of arithmetic. Formally, it has the advantages of being direct and of being applicable to transfinite numerosities. Young children, however, seem clearly to prefer (if not require) that decisions about the equivalence or nonequivalence of numerosities be based on the identity or nonidentity of their numerical representations rather than on the possibility of establishing a one-to-one correspondence between them. As we have already emphasized, the child's representations of numerosity derive from the counting procedure. Thus the child regards two sets as having equal numerosities if they yield the same cardinal numeron when counted (or perhaps, in some cases, if they both produce the same sequence when counted).

We assume that the judgment of equivalence rests on a counting procedure because of the striking increase in counting in the phase of the magic experiments when the equivalence of the arrays comes momentarily into question and when the child is asked to justify his judgments. In the part of the magic experiments that involves surreptitious transformations, children tend to count to determine whether or not number has changed (Gelman, 1972a). When the number has not changed, the child's assertion that the equivalence does or does not hold is accompanied by counting.

To sum up, the child's numerical reasoning principles recognize that between any two numerosities an equivalence relation may or may not hold. The practical decision about whether it holds or not rests on counting. If counting yields identical representations for the numerosities of the two sets, the sets are judged to satisfy the equivalence relation.

Ordering. Preschool children also seem to recognize that when numerosities do not satisfy a numerical equivalence relation they do satisfy a numerical ordering relation. In other words, given two numerosities x and y such that $x \neq y$, the child believes that either x is more than y or y is more than x. That is, the child believes that an ordering relation holds between x and y.

Evidence for this statement comes in part from the magic experiments in which the winning plate was transformed by either addition or subtraction of elements. In all of these studies the children not only recognized the resulting inequivalence but also gave clear evidence of recognizing what might be called the direction of inequivalence, that is, the ordering. The children's comments and repair behavior when items had been subtracted showed that they recognized that the numerosity of the transformed array was less than that of the original array. When items had been added the children understood that the numerosity of the transformed array was more than that of the original array (Gelman, 1972a, 1972b).

It should be noted that in the magic experiment the children were not making decisions based on direct comparisons of two perceptual arrays. The decisions concerned relations between a present array and a previously seen array. Furthermore, the magic studies showed that the decisions rested on an appreciation of a numerical ordering relation rather than of an ordering relation of some nonnumerical attribute, such as length or density of the sets. In the magic experiments that involved displacement rather than subtraction or addition, length and density were manipulated without disturbing the child's judgment that an equivalence relation rather than an ordering relation existed between the sets.

We are not assuming that the children use the terms *more* and *less*. Indeed, our subjects only rarely used these terms (Bullock and Gelman, 1977). Nevertheless, they clearly recognize the numerical ordering relation to which the terms refer. As the children seldom use these terms spontaneously in describing numerical ordering relations, it is not surprising that various studies of children's understanding of *more* and *less* (such as Clark, 1973; Palermo, 1973) are ambiguous or in conflict over whether the young child can deal with and represent order (compare Siegel, 1976).

Further evidence that the young child appreciates an ordering relation comes from a recent experiment by Bullock and Gelman (1977). Again the magic paradigm was employed. In Phase I, children were shown a one-mouse plate and a two-mouse plate. Half the subjects in each age group (2-, 3-, and 4-year-olds) were told that the one-mouse plate (less) was the winner, and half were told that the two-mouse plate (more) was the winner. There was no mention of number or quantity during this phase. Many children spontaneously identified the winner and loser on the basis of number, for example, "That loses, it has two; that wins, it has one." In Phase II the children were shown plates of three and four mice. The 3- and 4-year-olds clearly based their choice of a new winner on the quantitative relation—more or less—they had originally been reinforced for. In other words, those who had learned to consider the one-mouse plate the winner in Phase I called the three-mouse plate the winner in Phase II; likewise, those who had been told that the two-item array was the winner in Phase I chose the four-item array in Phase II. The 2-year-olds in this same experiment made random choices in Phase II and appeared to be unable to apply the ordering relation. A follow-up experiment made it clear, however, that 2-year-olds do *not* lack the ability to apply the ordering relation. To elicit this ability, it is necessary to provide them with clues that they should apply information gained in Phase I to the choices in Phase II. They seem to be unable to transfer the information without being cued. In other words, they lack the knowledge that they should transfer rather than the ability to apply the ordering relation (compare Baron, forthcoming).

Siegel (1974) showed that preschoolers can consistently respond to an ordering relationship between simultaneously present arrays. The experiment by Bullock and Gelman confirms Siegel's results. In addition, it shows that young children can use an ordering relation in an inferential manner. While in Phase I the children may have identified the winning and losing arrays by their absolute set size, in Phase II they made choices solely on the basis of the ordering relation.

In sum, when comparing small sets young children recognize that their numerosities are either equal or not. If the sets are not numerically equivalent, then the children reason as follows: If $x > y$, then the set with x items is more numerous; if $y > x$, then the set with y items is more numerous. The representations x and y appear to be obtained through a counting procedure.

To summarize formally, the young child recognizes that if two numerosities x and y do not satisfy the equivalence relation $x = y$ then they satisfy an ordering relation $>$, such that either $x > y$ or $y > x$.

Transitivity of the relations. The fact that many children spontaneously construct additional winner sets (Gelman and Tucker, 1975) might suggest that they regard the equivalence relation as transitive. In one of Gelman and Tucker's experiments, subjects started out with a winner plate containing two green mice and one soldier. In the surprise phase they encountered a plate with three green mice—which they judged the winner because it had three items while the other plate had two. At the end of this phase, the children were asked if they could fix the change in the winner. Almost all indicated that they could. Of interest are the children who asked for a soldier to substitute for the mouse (see D. S.'s protocol above). When these children were offered several objects, including two soldiers, they spontaneously left the three-mouse plate as it was and placed one soldier on the two-mouse plate—declaring that they now had two winners.

This behavior may be evidence that 3- and 4-year-old children regard the equivalence relation as transitive, or it may only indicate that judgments of equivalence form part of the cognitive principles with which children reason. A child may arrive at the decision that the newly created plate of three items is a winner in two ways, one that involves transitivity and one that does not. To demonstrate, let A_o represent the original winner array, A_t represent the array that underwent an identity substitution at the hands of the experimenter, and A_c represent the array that was transformed by the child into a winner array. The child may base his judgment that A_c is a winner on the following transitive reasoning: $A_t = A_o$ and $A_c = A_t$, hence $A_c = A_o$. Or the child may bypass the transitive reasoning and simply note directly that $A_c = A_o$. A general difficulty arises in any attempt to use experiments involving specific numerosities to determine whether the child's reasoning principles include transitivity. If one is working with some specified numerosity, say 3, then the question of transitivity becomes "If 3 = 3, is 3 = 3?" Clearly, it will be difficult to tell if a subject has used a conditional inference sequence as opposed to a simple judgment that 3 = 3. Transitivity becomes nontrivial only when one is dealing with *unspecified* numerosities, that is, only when numerosities are being considered algebraically. For example, the question of transitivity is not trivial when we ask, "If John has the same number of beans as Jane and Alice has the same number of beans as Jane, does Alice have the same number of beans as John?" The situation does not give an actual value for the three numerosities in question. Furthermore, the terms for the three numerosities (*John's beans, Jane's beans,* and *Alice's beans*) may or may not refer to the same numerosity. These terms—unlike numerons, numerlogs, and numerals—have the algebraic property that they can refer to different numerosities in different situations. As long as we test children's ability to judge equivalence and order only with specific numerosities, the question of transitivity is moot. The results of the magic experiment show that children's equivalence *judgments* are transitive when the numerosities involved are less than or equal to five. We are not aware of any unambiguous experimental evidence on whether young children's numerical *reasoning principles* include transitivity of the numerical equivalence relation. We must also leave unanswered the question of whether a principle of transitivity guides the inferences that children draw from the recognition of an ordering relation. For the children who participated in Bullock and Gelman's experiments clearly were working with specific numerosities.

The Operations

As just indicated, the magic experiments show that children have certain numerical reasoning principles that integrate their previous experience with their present experience. Numerical equivalence and order are important elements of these principles, elements that can organize comparisons between present and past experiences. The young child's reasoning about numerosities is not limited to the drawing of comparisons, however. The child interprets these comparisons by means of a scheme that categorizes possible transformations into number-relevant and number-irrelevant ones. The number-relevant transformations are further categorized into ones that decrease numerosity and ones that increase numerosity. The recognition that the numerosity of a given array is either more than, less than, or equivalent to the numerosity of the original array leads the child to postulate the intervention of a transformation. Judgments of equivalence go hand in hand with reference to manipulations that do not affect numerosity, and judgments of nonequivalence go along with the postulation of manipulations that do affect numerosity. Since the categorization of possible manipulations in this way plays much the same role in the child's reasoning that the operations play in formal treatments of arithmetic, we refer to these categories as *operators* (see Gelman, 1972a). In other words, as we explained in Chapter 5, operators are principles that specify the effects of manipulations.

Identity. As the magic experiments demonstrate, when children reason about numerosity they recognize the existence of a large class of transformations (manipulations) that can be performed on a set without altering the numerosity of the set. When called upon to explain unexpected spatial rearrangements, color changes, and item substitutions, they postulate transformations which have no effect on numerosity, such as lengthening and substitution. When probed the children will typically make statements showing that they realize that these transformations do not affect numerosity.

We will use *I* to symbolize the class of transformations that do not affect numerosity. The ability to recognize *I* transformations is mediated by an *I* operator in the child's class of operations. It is appropriate to note here that children do not always demonstrate the existence of such a classification scheme in their number reasoning. For example, in Piaget's well-known experiments (1952), children do not give evidence of using number-identity operators. However, in the magic experiments young children clearly do recognize the existence of number-identity transformations, that is, ones that do not alter number. Witness D. C.'s behavior after he confronts a row of five items that is considerably longer than he has come to expect.

D. C., age 4 years, 9 months.

What you did! What happened? *Look at that!* What? *Spreaded it out!* I did? *Yep.* Okay, is this the plate that wins a prize? *Yep.* How do you know that plate wins a prize? *Cause it has five.*

Or consider H. S.'s reactions when he confronts a change in the color of one of the mice on the three-item, winning array.

H. S., age 3 years, 11 months.

Aah! How did that get orange? What? *How did this get orange?* Did something happen to the plates? *Yeh. How did this get orange though?* I don't know. *Did you paint it orange?*

[Later in the questioning.] Is the plate different than it was at the beginning of the game? *Yeh.* How's it different? *Cause this is orange and these two are green.* What was it at the beginning of the game?

They were all green. Which plate wins? [Subject points to the 3-item plate.] How come that plate wins a prize? *Cause it has three.* Does it matter that one of them is orange? *Yeh, it does matter and it doesn't matter.*

To summarize, young children use a classification scheme that organizes operations into those that alter number and those that do not alter number. We symbolize the latter class of operations as I to indicate that it constitutes the class of identity operations. Formally speaking, the defining property of the class I of operations is $C[I(S_x)] = C(S_x)$, where S_x is a set of numerosity x, $C(\ )$ is the cardinal numeron obtained by counting the set specified within the parentheses (or brackets), and $I(\ )$ is the set produced when I operates on the set specified within the parentheses.

We do not know the limits of the class of identity operators available to the young child. In the adult, this class includes all operations except the few that are specifically assigned to the class of number-altering operations. We do know that transformations involving lengthening, shortening, or rotating a linear array or changing the color identity of an item in the array are all understood by young children in terms of identity operators. Thus the class of identity operators in young children is already quite extensive.

Addition. The young child's numerical reasoning scheme also includes operations that allow the child to deal with transformations that do alter numerosity. The first of these is addition. When young children confront an unexpected increase in numerosity, they postulate the intervention of addition (see Gelman, 1972a, 1972b). In other words, they state that something must have been added.

K. F., age 4 years, 3 months, participated in an experiment in which a three-item array was the winner and a four-item array the loser. In Phase II both arrays had four items.

Which one wins? [After K. F. has uncovered both arrays he stares.] *They're both four mice . . . How come that happened? . . . There's four mice but one of them . . . no . . . you see, one from that can came to this can and there . . . another one came to this can* [the can that had covered the 3-item array].

In order to explain unexpected increases in numerosity, the young child says that some set (containing one or more items) has been added to the original array. (Young children frequently postulate an agent of some kind for performing this operation.) Similarly, in the subtraction experiments, when children are asked how to repair the effect of subtraction, they say that some additional items should be joined with the items that are already there. In many cases, they add the items themselves. Thus at the behavioral level the child's discussion of and performance of addition always involves the uniting of disjoint sets. Therefore, we assume that conceptually the young child regards addition as the uniting of disjoint sets.

To obtain a representation of the numerosity of the union, the young child uses the same procedure that he uses to obtain a representation of any other numerosity—he counts. Previous investigators have documented the fact that the young child initially evaluates the outcome of addition by counting the set that results from uniting two disjoint sets (for example, Beckmann, 1924; Brownell, 1941; Ginsburg, 1977; Ilg and Ames, 1951; Reiss, 1943; Wang, Resnick, and Boozer, 1971). The following protocol is typical of early addition.

> Father: How much is two and three?
>
> 3-year-old son: [Holding up his right hand, he counts.] *One, two, three. That's three.* [Keeping three fingers on his right hand upright, he holds up a closed left hand and raises two fingers, one after the other.] *That's two.* [Then he puts together the fingers that are upright and counts them.] *One, two, three, four, five.*

This protocol involves a child whose father was trying to teach him to add "in his head," that is, without observable counting. Since such training is common, both at home and at school, most children (including some preschoolers) eventually know their "number facts" (Brownell, 1941). They can state from memory the outcome of adding any pair of small numbers. In between the use of overt counting of the union and the immediate production of the answer from memory, one frequently sees a more sophisticated counting strategy: The child begins his count with the cardinal number of one of the sets composing the union, and then adds by counting up from there (see, for example, Ginsburg, 1977). Thus a child may add eight and three by counting "eight—nine, ten, eleven." Note that this method involves a recursive use of the counting procedure: The child must be counting the iterations in the counting sequence. Otherwise how would the child know to stop three counts beyond eight? Groen's studies (1967) of the time it takes children of school age to add suggest that many of them continue to apply this strategy covertly. Although the older children may not count aloud, the pattern of their reaction time in addition problems strongly suggests that they start with one of the numbers and count up from there.

Given that the child defines the numerosities of sets in terms of the outcome of the counting process, it is hardly surprising that he evaluates the consequences of addition by counting. And the fact that the young child uses counting to evaluate the effects of addition more or less forces the addition operation to be defined (at an early point in development) only for the union of disjoint sets. Recall that counting involves tagging each item once and only once. This in turn necessitates a step-by-step partitioning of the counted items from the to-be-counted items. This procedure would not work if nondisjoint sets were to be combined under an addition operation whose outcome was to be evaluated by counting. Imagine that you have a round block and a square block and you ask a child to add the round blocks to all the blocks. If the child begins by counting the round block, he thereby partitions it into the group of already counted blocks. But the round block also belongs to the set of all blocks and therefore should be on the other side of the partition, among the group of blocks still to be counted. Clearly the counting process, as we have described it, will be stymied by the request to add two nondisjoint sets.

More generally, we believe that the child assumes that tasks requiring reasoning about numerosity make sense only when there are disjoint sets involved. The work of Piaget seems to support this conjecture. Piaget has found that children do very poorly when asked to judge the relative numerosity of a set and a proper subset of that set. For example, when young children are asked "Which is more, the flowers [the set] or the tulips [a proper subset]," they give the wrong answer. They say the tulips have more—apparently because they compare the subset of tulips with the other disjoint subset of flowers, the roses. And indeed in these experiments the subset of tulips *is* more numerous than the only other set in the situation that is disjoint with respect to the tulips. Of course, neither subset of flowers is disjoint with respect to the set of flowers. The explanation, then, for the child's difficulty in such Piagetian experiments is that the child assumes that reasoning questions about numerosity must refer to disjoint sets. A similar argument was developed independently by Wilkinson (1976), who finds that the young child's counting strategies lead him to err on class-inclusion problems.

To sum up formally, the operations, A, that the child recognizes as addition have the following defining properties: They can be applied to two sets, S_x and S_y, if and only if $S_x \cap S_y = \emptyset$, and when applied, $C[A(S_x,S_y)] = C(S_x \cup S_y)$; where $S_x \cap S_y = \emptyset$ means that the two sets are disjoint, $A(\ ,\ )$ symbolizes the set produced by the addition operation, $S_x \cup S_y$ symbolizes the set-theoretic union of the two sets, and $C[\ ]$ symbolizes the number (numeron) obtained by counting the set in brackets.

Subtraction. Just as our magic experiments show that children know the effects of addition, they also provide evidence that young children use another number-altering operation: subtraction.

V. B., age 4 years, 4 months, participated in the subtraction condition of an experiment involving a five-mouse plate and a three-mouse plate in Phase I. The five-mouse plate was the winner and was changed to a three-mouse plate in Phase II.

[Phase I.] Why win? *Cause there's one, two, three, four, five.* Why lose? *Cause one, two, three.*

[Phase II.] [First V. B. uncovers the three-mouse plate.] Win? *No . . . three mouses.* Okay, which plate wins? [V. B. points to remaining can and lifts it.] Win? *Wait! There's one, two, three.* Is that the plate that wins? *No.* Why? *Because it has three. It has three!* What happened? *Must have disappeared!* What? *The other mouses?* Where did they disappear from? *One was here and one was here.* [She points to spaces on the nontransformed plate.] How many now? *One, two, three.* How many at the beginning of the game? *There was one there, one there, one there, one there, one there.* How many? *Five—this one is three not but before it was five.* What would you need to fix the game? *I'm not really sure because my brother is real big and he could tell.* What do you think he would need? *Well, I don't know . . . some things come back.* [Experimenter hands V. B. some objects, including four mice. V. B. puts all four mice on one plate.] *There. Now there's one, two, three, four, five, six, seven! No . . . I'll take these* [points to two] *off and we'll see how many.* [V. B. removes one and counts.] *One, two, three, four, five, no—one, two, three, four. Uh . . . there were five, right?* Right. *I'll put this one here* [on table], *and then we'll see how many there is now.* [V. B. takes one off and counts.] *One, two, three, four, five. Five! Five.*

As is evident in this protocol, the young child regards subtraction as the removal of items from a set. Notice the sequence of reasoning in the protocol. The child had clearly stored in memory a representation of the numerosity of the winner plate. When confronted with the altered array, she obtained a representation of its numerosity. She compared the numerosity of the altered array with the stored representation of the numerosity of the winner plate. She noted that the equivalence relation did not hold between these two representations of numerosity; rather, the ordering relation held. From the direction of the order relation, namely, that the numerosity of the new set was less than that of the original set, she concluded that an item or items had been taken away.

To sum up formally, the operations, M, that the child recognizes as subtraction have the following defining properties: They can be applied to two sets, S_x and S_y, if and only if $S_x \subset S_y$, and, when applied, $C[M(S_x,S_y)] = C(S_y - S_x)$; where $S_x \subset S_y$ means that S_x is a proper subset of S_y, $M(\ ,\ )$ symbolizes the set produced by the removal of the subset, and $S_y - S_x$ is the set theoretic difference of the two sets, and $C[\ ]$ again is the representation of numerosity obtained by counting the set in brackets.

Solvability

In the magic experiments children encountered sets whose numerosity was either more (the addition experiments) or less (the subtraction experiments) than the numerosity they expected. We have already mentioned that children reliably indicated the direction of the discrepancy and the operation that caused the discrepancy. In addition, the children knew how to eliminate the discrepancy. They made cogent suggestions (see V. B.'s protocol above) and were typically able to carry them out if offered suitable props. In other words, when confronted with the discrepancy between an actual numerosity, n, and an expected numerosity, m, they showed that they knew that m could be converted into n by either addition or subtraction. As we have already emphasized in the discussion of the ordering relation, the children reliably applied the appropriate operation. When m was less than n, they specified addition; when m was greater than n, they specified subtraction. When the difference between n and m was equal to one, the children did more than apply the appropriate operation; they also specified the number to be added or subtracted. This statement, as always, applies only when the numerosities of n and m are both small (less than or equal to four). As the difference between n and m became greater than one, the children reliably indicated that the number to be added or subtracted was greater than one, but they became less precise about the exact value of that number. This evidence from a recent magic study by Gelman (1977) is summarized here again because it is central to our argument about how young children go about undoing additions or subtractions that involve more than one item.

Summary of the magic experiment. In Phase I of the [(3 vs. 5) – 2] magic experiment, 54 children (30 3-year-olds and 24 4-year-olds) were tested. A plate with three green mice in a linear row was the loser. A plate with five green mice in a linear row was identified as the winner. As in other experiments, the difference in number was redundant to a difference in either length or density. Six of the 3-year-olds were dropped after Phase I for failing to reach the criterion of five out of six correct identifications; these children could not keep straight which was the winner and which was the loser. Dropping them left an equal number of children in each age group.

In Phase II, children in the subtraction conditions encountered a winning plate that had three mice, two fewer than expected. Children in the displacement conditions encountered a shortened or lengthened row. For all remaining details of the procedure, see Chapter 8.

The Phase I results compare to those of our other studies. Thirty-four of the children made no identification errors at all. For the eight 3-year-olds who erred at least once, the mean number of errors was 1.4; for the six 4-year-olds who erred, the mean error score was 1.5. As in previous experiments, the children had a striking tendency to define the winner and loser spontaneously in terms of their absolute number. All but six children (three in each age group) talked about numerosity. Clearly Phase I established an expectancy for number in the children. How did they react to the unexpected changes in Phase II?

In most ways, these children responded to the Phase II events just like children who participated in other displacement-versus-subtraction experiments. Children in the displacement condition treated the effects of this transformation as irrelevant: They said they still won because the number of items was as expected. If asked about the change in length they were able to state which transformation had produced it. Children in the subtraction condition treated the surreptitious change in number as a violation of their expectation that the five-mouse plate would win. The altered array did not win because it had only three items. The change in number produced considerable surprise and search behavior, and the children assumed that somehow items had been removed from the display (see V. B.'s protocol above). Thus, as we expected, the children revealed an ability to make inferences about the operations that could produce the transformations they encountered.

As we mentioned earlier, the results of the [(3 vs. 5) – 2] experiment differed from those in which the subtraction or addition involved only one item. When one item was added or removed, children were precise about the size of the deviation. Furthermore, they were precise about the number of items—one—that needed to be added or subtracted to "fix" the game. In the present experiment, where they expected a set of five items and encountered one of only three items, they were not nearly as precise on these counts. They knew that some items had been removed and generally gave evidence of knowing that more than one item was missing. In various ways, 26 of the 32 children talked of more than one missing item: "They gone." "Some came out." "Has to be some more." Yet only six children could state that terms like *they, some,* and *some more* had the specific numerical reference of *two.* In other words, the ability to solve, in their heads, for the specific difference between the expected numerosity and the actual numerosity seems to be poorly developed in preschool children. The way they "fix" the game, however, clearly shows that they recognize that *in principle* it is possible to solve for the difference.

When asked how to fix the game, all but two children indicated the need to add some items. When given four mice, only four children took just two of them. The rest began by taking one, three, or four and placing that number of mice on a display. What followed was a sequence of counting, adding or subtracting, counting again, and so on—much like that in V. B.'s protocol. Eventually 11 of the 16 4-year-olds produced a five-item array and declared it to be like the original. Only four 3-year-olds met this criterion. Despite the fact that many children did not end up with a five-mouse plate, all but three ended up with a winning array that had more than three items, arrays of from four to seven items. Thus the children knew that items should be added but did not necessarily know exactly how many items were needed to repair the game.

The principle of solvability. What does the [(3 vs. 5) − 2] experiment add to our understanding of the young child's arithmetic principles? It shows that the principle that guided the young child's repair behavior in the initial magic experiments was not limited to differences of only one. Young children are not very good at specifying larger differences, but they are able to indicate that the difference is greater than one. Further, they know how to begin to remove the difference. Almost all of the children in the subtraction condition knew that they should add some items. Not knowing the exact number, they typically proceeded through a trial-and-error sequence of counting and adding or subtracting.

We hesitate to take these results as evidence for granting young children a precise concept of the inverse. Still, much in their behavior warrants the postulation of some principle of reversibility, that is, some principle that leads the child to recognize that addition is what undoes the effect of subtraction and to attempt to alter the arrays in a systematic fashion. What is the simplest principle that explains this repair behavior? We think it is a *principle of solvability,* or the "you can get there from here" principle. Put more formally, this principle states that given two sets, S_x and S_y, such that $C(S_x) < C(S_y)$, there exists a set S_a such that $C[A(S_x,S_a)] = C(S_y)$, and there exists a set S_m such that $S_m \subset S_y$ and $C[M(S_m,S_y)] = C(S_x)$.

We assume that here, as elsewhere, the solvability principle is put into practice through the application of an algorithm involving the counting procedure. How the child does this depends a great deal on how he has represented the numerosity of the winner set. If he has a unitary representation of the winner set's numerosity, that is, if the numerosity is represented in his memory only by a cardinal numeron, then solving for the difference between the expected and actual numerosities is likely to involve either a recursive use of the counting procedure or the creation of sets of mental entities that match his stored representation of the numerosity. Assume that the child has a unitary representation of the numerosity of the winner together with a representation of the numerosity that now confronts him. Observing the direction of the ordering relation between these two representations tells him which operation (addition or subtraction) he must apply to eliminate the difference. However, the children clearly do more than this. They calculate some representation of the numerosity of the difference. They clearly distinguish, for example, between a difference of one and a difference of two.

All of our evidence suggests that the algorithms children use to calculate numerosities involve counting. How can the counting procedure be employed to obtain a representation of the numerosity of the difference between the expected and the actual arrays? One way is to use the counting procedure recursively. It is possible to begin with (or count up to) the representation of the smaller numerosity and count from there to the representation of the larger numerosity, keeping track of the number of steps in the count. Since keeping track of the number of steps means counting the iterations in one's counting procedure, this algorithm employs the counting procedure recursively. The other possibility is that the child uses his representation of the larger numerosity to guide the creation of a homomorphic set of mental entities, that is, a mental set of some kind whose numerosity is equal to the numerosity of the larger of the two physical sets. He can then count off and remove a subset of these mental entities equal in numerosity to the numerosity of the smaller of the two physical sets. With this subset removed, a representation of the difference can be obtained by simply counting the remaining mental entities. This method does not involve a recursive use of the counting procedure.

Alternatively, the child's representation of the numerosity of the winning array may be not unitary but rather homomorphic to the physical array. This is the case if the child stores the entire count sequence rather than just the cardinal numeron. In this case the child does not have to create a set of mental entities; he can work directly on the set contained in his representation of the larger array. He must count off and remove from this mental set a mental subset equal in numerosity to the numerosity of the smaller physical array and then count the remainder.

In any event, it appears that solving for the numerosity of the difference between two representations of unequal numerosities is not simple. It is surprising that children as young as 3 do it at all and not surprising that they are rather poor at getting the precise answer when that answer is greater than one.

Although we have no evidence that allows us to choose among the alternative difference-solving algorithms, we should point out that it might be possible to test whether a recursive algorithm is being used. The recursive use of the counting procedure should greatly compound the coordination problem posed by the counting procedure. We have already seen that the coordination is more likely to break down with longer count sequences. Furthermore, when the recursive algorithm is used to solve for the difference, both the first count sequence and the recursive second count sequence need be only as long as the difference. Thus if children do use the counting procedure recursively the error rate and uncertainty in solving for a difference should be more closely related to the size of the difference than to the absolute size of the sets. For the two nonrecursive methods, difficulty and uncertainty should be at least as closely related to the absolute numerosities involved as to their difference.

The way young children behave in the magic experiments has led us to postulate a set of arithmetic reasoning principles that young children use when they reason about small numbers. These principles make possible inferences about numerical equality and ordering relations, transformations that do or do not alter set size, and how to reverse the effects of addition and subtraction. To be sure, these principles are less sophisticated than those we attribute to older children and adults. Nevertheless they *are* reasoning principles. The evidence indicates that young children have some ability to reason about numbers. The question that remains is how such principles underlie the development of more complex and extensive principles of arithmetic reasoning.

Chapter Eleven

Formal Arithmetic and the Young Child's Understanding of Number

Many features of formal arithmetic systems are clearly present in some form in the numerical reasoning of the preschooler. Analyzing both the similarities and the differences between the preschooler's reasoning and formal arithmetic reasoning gives us a clearer picture of the child's number concepts.

Formal arithmetic is defined by the so-called laws of arithmetic. There are several slightly different versions of these laws—several constitutions, so to speak. We have chosen the one put forward by Konrad Knopp (1952), because it seems to us to be an unusually natural one, and because it is framed by some other material we want to use. Knopp provides a complete set of laws (except for the lack of a technically important principle, the Archimedean principle). These laws form the starting point for most higher mathematics.

We might have chosen an incomplete set of laws—the so-called Peano axioms. Working with this incomplete set of axioms, one can, through a series of definitions, arrive at a complete set of laws similar to those of Knopp. The laws given by Peano are interesting in that they apply only to the natural numbers, 1, 2, 3, and so on. However, the Peano axioms do not "look like" old familiar arithmetic to most people. Hence, they do not serve well as a backdrop against which to perceive the special properties of the child's reasoning.

The theory of groups provides a framework for comparing and contrasting the child's system of reasoning and the formal arithmetic system. This theory describes systems that have features in common with the system defined by the laws of arithmetic. As a prelude to making our comparison, we state the laws that define the modern system of arithmetic, then comment on the relation between these laws and the popular concept of what constitutes a number, and finally describe the group-theoretic structure of the system.

In comparing "the preschooler's system" and modern arithmetic, we have in mind the advanced preschooler. The extent to which the system we grant the preschooler in this comparison may have to be qualified for 2- or 3-year-olds will be taken up in Chapter 12. Also, as will become apparent, the modern system of arithmetic should by no means be viewed as the "adult" system. It is all but certain that the reasoning of many adults lacks some essential features of standard formal arithmetic. Many adults, for example, have never fully come to terms with negative numbers. They may or may not remember the rules they learned for handling these alien entities when they were in school. If they do remember, they may nonetheless sympathize with the schoolboy jingle reported by W. H. Auden: "Minus times minus is plus/The reason for this we need not discuss." Even adults who feel comfortable with negative numbers may be unsettled by paradoxes that bothered mathematicians as distinguished as Pascal and Leibnitz: How, for example, can the ratio $1/-1$, which is the ratio of a larger number to a smaller number, be equal to the inverse ratio $-1/1$, which is the ratio of the smaller number to the larger number? Whatever intellectual discomfort negative numbers create, they are an essential part of modern formal arithmetic. Take them away and the formal structure evaporates. We mention the essential presence of these discomfiting "less-than-nothings" in order to emphasize that modern arithmetic cannot be regarded as a model of adult numerical reasoning. The modern formalism is an elegant but exotic system. It does not flourish in the untutored mind.

We compare the preschooler's arithmetic principles to the modern formal system of arithmetic in order to highlight various aspects of the preschooler's principles. The modern formalism is the only available backdrop, because no one, as far as we know, has formalized what might be called the layman's system of arithmetic.

The Laws of Arithmetic and the Definition of Number

The formal description of modern arithmetic is a product of late-nineteenth-century and early-twentieth-century mathematics. Its emergence was accompanied by a profound shift in mathematicians' views regarding the relation between the laws of arithmetic and the definition of what constitutes a number. The premodern view of this relation was roughly as follows: The knowledge of what is and is not a number is the mental bedrock upon which arithmetic is built. The laws of arithmetic are dictated by this knowledge of what a number is. The concept of number is primary; from this primary intuition flow the rules for what you can and cannot do with numbers. This view is to some extent captured by the pronouncement of a distinguished turn-of-the-century mathematician: God made the integers/All else is the work of man (Leopold Kronecker, cited in Kline, 1972, p. 979). The modern formal approach reverses the old relation between the definition or "raw concept" of number and the laws of arithmetic. It argues that the laws of arithmetic are the bedrock. These laws determine what is and is not a number. A number, in the modern view, is any abstract entity, no matter how bizarre or repellent to one's numerical common sense, that can be shown to behave in accord with the laws of arithmetic. For example, in the modern view the string of symbols $(2 + 3\sqrt{-1})$ is a single number (a complex number), because it and other strings of symbols like it can be manipulated as single entities in a way that conforms with almost all of the rules of arithmetic.

The modern view sharply distinguishes between the abstract entities called numbers and the concrete realities of the physical world, which the abstract entities may to some extent represent. The *number two* is an abstract entity that can be manipulated in accord with the laws of arithmetic. *Numerosity two* is an aspect of many collections of objects, an aspect that we can represent by the number two. The number two may also represent aspects of the world other than numerosity, such as the length of a line or the speed of a tricycle. Some numbers (the negative numbers) cannot represent concrete numerosities. They can only represent "fictional" numerosities, such as the balance of dollars in an overdrawn checking account. Still other numbers (the irrationals) cannot represent numerosity at all; they can only represent magnitudes, and other continuous quantities. The distinction between the abstract number two and an aspect of reality (numerosity two) that can be represented by that number leads one to ask how the abstraction can be coupled to the reality. How do we link numerical representations to what they represent? By what principles do we determine which number represents which numerosity? These questions lead in turn to the viewpoint that underlies our analysis of the child's understanding of number: namely, that the principles by which one abstracts number from reality are distinct from the principles by which one reasons about number. In distinguishing between the principles of numerical reasoning in the preschooler and the principles by which the preschooler abstracts number from reality, we have been influenced by the parallel distinction within mathematics. The question of which numbers should represent which numerosities and the related question of whether two or more numerosities are numerically equivalent are no concern of arithmetic. These questions and the principles relevant to them are addressed in other branches of mathematics—mathematical logic and the theory of measurement. Arithmetic proper does not consider what numbers represent nor the principles by which one establishes a correspondence between numbers and numerosities (or numbers and any other aspect of reality).

In presenting the laws of arithmetic, we follow Konrad Knopp's development of the number system (1952). Knopp states the modern formalist view of arithmetic clearly and forcefully. He begins by mentioning the rational numbers. The rational numbers are, roughly speaking, the numbers that any child past the sixth grade is familiar with, namely, the natural numbers (1, 2, 3, . . .), plus the number 0, plus the negative numbers, plus fractions. Using the word *numbers* to refer only to the rational numbers, and denoting these numbers by lowercase roman letters, Knopp sets down the following laws:[1]

> I. FUNDAMENTAL LAWS OF EQUALITY AND ORDER
> 1. The set of numbers is an ordered set; i.e., if a and b are any numbers, they satisfy one, and only one, of the relations
> $a < b, \quad a = b, \quad a > b.$
> This order obeys these additional laws:
> 2. $a = a$ for every number a. [Reflexivity of equality.]
> 3. $a = b$ implies $b = a$. [Symmetry of equality.]
> 4. If $a = b$ and $b = c$, then $a = c$. [Transitivity of equality.]
> 5. If $a \leq b$ and $b < c$, or if $a < b$ and $b \leq c$, then $a < c$. [Transitivity of order.]
>
> All numbers which are greater than zero are called positive, all numbers which are less than zero are called negative. If a number is equal to zero, we also say that it "vanishes."
>
> II. FUNDAMENTAL LAWS OF ADDITION
> 1. Every pair of numbers a and b can be added; the symbol $(a + b)$ or $a + b$ always represents a definite number, the sum of a and b. [Closure of numbers under addition.] This formation of sums obeys these laws:
> 2. If $a = a'$ and $b = b'$, then $a + b = a' + b'$. ("If equals are added to equals, the sums are equal.")
> 3. $a + b = b + a$. (Commutative law.)
> 4. $(a + b) + c = a + (b + c)$. (Associative law.)
> 5. $a < b$ implies $a + c < b + c$. (Monotonic law.)

1. The terms in brackets are our verbal labels for Knopp's laws.

> III. FUNDAMENTAL LAW OF SUBTRACTION
>
> The inverse of addition can always be performed; i.e., if a and b are any numbers, there exists a number x such that $a + x = b$.
>
> The number x thus determined is called the difference of b and a, and is denoted by $(b - a)$. (Knopp, 1952, pp. 3–4.)

After completing the presentation of the fundamental laws by giving the laws of multiplication and division, Knopp points out that once we have established the validity of these laws all other rules for dealing with numbers can be derived directly from them without making any use of the fact that the symbols a, b, x, y, and so on denote rational numbers, that is to the kind of thing one had in mind in laying down the laws of arithmetic in the first place. This realization leads Knopp to a dramatic conclusion:

> From the important fact that the *meaning* of the literal symbols need not be considered at all . . . , there results immediately the following extraordinarily significant consequence: If one has any other entities whatsoever besides the rational numbers . . . *but which obey the same fundamental laws,* one can operate with them as with the rational numbers, according to exactly the same rules. Every system of objects for which this is true is called a *number system,* because, to put the matter baldly, it is customary to call all those objects *numbers* with which one can operate according to the fundamental laws we have listed (p. 5).

In other words, it does not matter who the players are so long as they are constrained to play by the rules. This is the essence of the formal approach to defining numbers (and other mathematical entities). Knopp's particular reason for emphasizing the point is to prepare the reader to regard as numbers entities that are unlike what most laymen regard as numbers.

The things commonly regarded as numbers are, first and foremost, the *natural* numbers, that is, the numbers one gets by counting: the positive integers, 1, 2, 3, and so on (called in German the *Zahlen,* or counting numbers). The other numbers we are used to dealing with are all generated from the natural numbers through the application of the inverse operations of subtraction and division. Zero is generated by calling the result of subtracting 3 from 3 (or 4 from 4, and so on) a number in its own right. The practice of regarding this entity as a number was not established in Western mathematics until the Renaissance (Kline, 1972). The negative numbers are generated by subtracting positive integers from smaller positive integers. This procedure and the resulting "numbers" were not regarded as kosher until very recently. The law that "the inverse of addition (that is, subtraction) can always be performed" was not an accepted law of arithmetic until the seventeenth century. Even then, it was accepted only in the most advanced mathematical circles. Euler, one of the half-dozen greatest mathematicians of all times, believed that negative numbers were greater than infinity. The distinguished nineteenth-century mathematician, DeMorgan, argued that subtracting a number from zero was inconceivable (Kline, 1972). The positive fractions are generated by dividing one natural number by another natural number, say 1 by 4. By subtracting the positive fractions from, say, zero, one gets the negative fractions. This exhausts the list of entities that the average post-Renaissance man has come to regard as numbers. These numbers are called the *rational* numbers.

Modern mathematics, however, requires that more exotic entities be considered numbers. The modern formal development of arithmetic has been shaped by this demand. For example, it is essential to modern mathematics that $\sqrt{2}$ be a number. But there exists no natural number, nor any number that can be generated from natural numbers through the unfettered use of subtraction or division, that, when multiplied times itself, is equal to two. This uncomfortable fact was first proved by Euclid. It was an unresolved problem at the core of mathematics for two millennia thereafter. At the end of the nineteenth century the problem was resolved by devising for the first time a means of rigorously defining the *irrational* numbers. The $\sqrt{2}$ belongs to this class of numbers. These recently defined numbers are irrational in that, unlike the rational numbers, they cannot be generated by applying division or subtraction to the natural numbers.

We believe that in the mind of the child the mathematician's terminology takes on a literal significance. The numbers generated by counting *are* the natural numbers—indeed the only numbers—that the child recognizes. Numbers that can in no simple way be derived from these natural numbers would seem irrational to the child, that is, foreign to his system of numerical reasoning. The normal preschool child's numerical reasoning appears closely tied to the procedure that generates the mental entities that he manipulates when he reasons numerically. And that procedure is counting. Thus, the child's arithmetic system is strongly shaped by the mental entities with which it deals, namely, the representations of numerosity that may be obtained by counting. The child's arithmetic system departs most noticeably from the modern formal system at just those points that require recognizing as numbers mental entities that cannot be obtained by simple counting. Nonetheless, at most such points, the child does employ principles that could lead on to the generation and use of such "noncounting" numbers.

Group-Theoretic Description of Arithmetic

Since we have not determined whether the operations of multiplication and division figure in the preschooler's numerical reasoning, we consider the modern system of arithmetic as it appears when shorn of these operations. When thus trimmed, modern arithmetic can be characterized as a *commutative* (*abelian*) *group* on which an ordering relation is also defined. An abelian group is a set for which an addition operation is determined in some way, that is, for which a procedure is specified that determines for each pair, a, b, of elements in the set another element, c, also in the set, called their sum. The procedure must obey the following rules:

1. $(a + b) + c = a + (b + c)$. (The associative rule.)
2. $a + b = b + a$. (The commutative rule.)
3. For each two elements, a, b, there exists an element c such that $a + c = b$.

Taking ordinary arithmetic addition as our addition operation and allowing it to operate on a set of elements consisting of the positive and negative integers and the integer 0 yields an example of an abelian group. The operation of adding any two integers yields another element in the set, namely, the integer that is their sum. This addition is commutative: It does not matter which integer is added to which. It is also associative: The order in which three or more integers are added does not matter. Knopp's fundamental law of subtraction also holds: Given two integers, there is a third integer equal to the difference between them. For example, given the numbers 3 and 2, there exists a number, namely -1, such that $3 + -1 = 2$. This group-theoretic structure (that is, this commutative group) is not the structure we call arithmetic, because an ordering relation does not appear anywhere in it. Nowhere in the course of combining pairs of integers by addition so as to produce other integers is it necessary to consider whether one integer is greater than or less than another. Indeed, the structure we have so far described applies equally well to nonnumerical systems in which the question "Is that element greater than this one?" does not make sense.

Consider, for example, a clock with a single hand that moves in steps around the dial. Assume, for the sake of concreteness, that the pointer completely circles the dial in 10 steps. We can regard the 10 possible positions as the elements of a set. And we can make up an addition operation by specifying the following procedure: Consider straight up as the identity position of the pointer. Any other position of the pointer is then a characteristic number of clockwise steps away from the identity position. The first position to the right of straight up is one step away, the first position to the left is nine steps away, and so forth. Our addition procedure says that in order to add any two positions, you simply add the number of steps required to get from the identity to one of the positions to the number of steps required to get from the identity to the other position (stepping in the clockwise direction only). This process determines a position, which we shall call the sum of the other two positions. The sum of the second and fifth positions is the seventh position. The sum of the second and eighth positions is the identity position. The sum of the seventh and eighth positions is the fifth position. The fourth position, when added to itself, yields the eighth position, and so on.

This 10-step circular system is an abelian group. Any two positions can be added to produce a position. It does not matter which of the two positions is added to which (commutativity); nor, in adding three positions, does it matter which two are added first (associativity). Finally, for any two positions there exists a position that when added to one of them yields the other. This statement holds even for the interesting (and formally important) case in which our "any two" positions are one and the same. Suppose some skeptic asks "Any two positions? How about position seven and position seven? Is there a position you can add to seven in order to get seven?" The answer is "Yes, the identity position." The identity position plays the role of zero in the number system. Adding the identity position to any position yields that position. Algebraically, $a + 0 = a$, where 0 can now stand either for the integer 0 or for the identity position on our clock. Clearly, the positions on our clock together with our made-up addition procedure constitute an abelian group.

In our circular abelian group it does not make sense to ask whether one position is greater than or less than another. In fact, any simple attempt to specify such an ordering relation for the positions of the clock will quickly run into confusion. If we consider the seventh position to be greater than the second position, we will have to confront the paradox that adding the fourth position to each will result in sums with reversed order. Position a is greater than position b, but the sum of a and c is less than the sum of b and c! The ordering is not monotonic with respect to addition; or, putting the matter algebraically, it is not always true that if $a > b$, then $a + c > b + c$. It would be hard to define an ordering relation for the positions that behaves the way we want it to behave when we allow our addition operation to operate. The combination of our addition operation and our ordering relation will not satisfy the monotonic law.

The above example was designed, first, to illustrate the nature of abelian groups, and second, to show that formal arithmetic is an abelian group upon which an ordering relation is also defined. The ordering relation is governed by two rules:

1. If $a \neq b$, then either $a > b$ or $b > a$, but not both. (Antisymmetry.)
2. The ordering is transitive: if $a < b$ and $b < c$, then $a < c$.

Finally, in order to complete the formal structure that is arithmetic we need two rules that coordinate the addition operation with the equivalence and ordering relations:

1. If $a = a'$ and $b = b'$, then $a + b = a' + b'$. (When equals are added to equals the results are equal.)
2. If $a < b$ then $a + c < b + c$. (The monotonic law.)

Comparison to the Child's Numerical Reasoning

One thing to note in comparing modern arithmetic to the child's scheme is that in the formal scheme the subtraction operation is not a separate operation. This elimination of subtraction as an operation in its own right was made possible by the introduction of negative numbers. Once we regard negative two as a number, we can regard the operation of *subtracting* two from three as the operation of *adding* negative two to three. This is what Knopp has done in his formulation of the fundamental law of subtraction. He has defined subtraction in terms of addition. His statement that if a and b are any numbers, there exists a number x such that $a + x = b$ is equivalent to specifying another operation, subtraction, which satisfies two conditions:

1. The operation can always be performed, that is, given any two numbers, a and b, there exist numbers, x and $-x$, such that $a - b = x$ and $b - a = -x$.
2. The numbers x and $-x$ must always have the following properties: If $a - b = x$, then $a = b + x$; and if $b - a = -x$, then $b = a + -x$.

Point 2 in this definition of subtraction expresses an interrelation between subtracting a from b (to get x) and adding the resulting x to b. Adding x to b must produce a. Because of this interrelation it is more elegant (economical) not to define addition and subtraction as separate operations but rather to subsume one under a rule governing the other. The economy and elegance are formal matters. They do not mean that there is no such thing as a subtraction operation in formal arithmetic. There is; but in formal descriptions of arithmetic it is subsumed under the rules defining the addition operation.

The rules defining (or subsuming) subtraction require that zero and the negative numbers be regarded as numbers. If zero is not a number, then when $a = b$ there does not exist a number x such that $a + x = b$. If the negative numbers are not numbers, then when $a > b$ there exists no x such that $a + x = b$. Historically, the desire to have the subtraction operation be as generally applicable as the addition operation (that is, applicable to *any* two numbers) was part of the motivation for admitting zero and the negative numbers to the realm of the legitimate numbers (Kline, 1972; Gardner, 1977).

The entry of zero and negative numbers into the system of arithmetic constitutes a historical example of assimilation and accommodation. The assimilation of new numbers led to an accommodation (change) in the structure of the system into which those numbers had been assimilated.

One must grasp the manner in which the laws of arithmetic have changed and grown in response to the introduction of new numbers in order really to appreciate the reasons behind the law of signs. One "does not discuss" the reason why minus times minus is plus with a sixth grader because to grasp the reason fully requires the kind of abstract and purely formal view of the structure of arithmetic that is difficult to instill in the child's teacher let alone in the child. Minus times minus is plus because only under that rule do the assimilated numbers (the negative numbers) behave like the old numbers. The distributive law governs the relation between addition and multiplication, saying that $a(b + c) = ab + ac$. By letting $a = -1$, $b = -2$, and $c = +3$, the reader may see that the distributive law will not hold unless minus times minus is plus. The law of signs is an extension or elaboration of the laws of arithmetic that is necessary because of the assimilation of new numerical entities.

Both the treatment of subtraction and the law of signs are historical instances of conceptual accommodation in response to the assimilation of new elements into the set of elements that the conceptual scheme deals with. Similarly, the rules that govern the child's numerical reasoning are influenced by what the child regards as belonging to the domain of mental entities that are to be reasoned about numerically. The mental entities to which the child's numerical reasoning principles apply are his representations of numerosity. Because his representations of numerosity derive from a counting procedure, he has no numerical representations corresponding to zero and the negative numbers. The rules that govern his numerical operations therefore cannot all be subsumed under the rules defining an addition operation.

The child's numerical reasoning involves two or possibly three operations that supplement the addition operation. First, there is the identity operation. Where formal arithmetic states that $x + 0 = x$, the child reasons that $C[I(S_x)] = C(S_x)$.

Second, there is the subtraction operation. The formal rule of subtraction can be expressed as follows: For all numbers a and b, there exists a number x such that $a + x = b$. The child's rule governing subtraction is as follows: For all sets S_x and S_y, such that $S_x \subset S_y$, $C[M(S_x,S_y)] = C(S_y - S_x)$.

Third, the young child has a limited solvability principle. He believes that a lesser numerosity may be made equivalent to a greater numerosity by means of the addition operation and that a greater numerosity may be made equivalent to a lesser numerosity by means of the subtraction operation. Embedded in this belief is the belief that addition always increases numerosity and subtraction always decreases numerosity. In other words, this principle embodies his notion of how the order relation relates to the addition and subtraction operations. The child's limited solvability principle is akin to the monotonic law in formal arithmetic.

In formal arithmetic, the principle might be stated like this: Given that $a < b$, there exist c and $-c$ such that $c + -c = 0$, $a + c = b$, and $b + -c = a$.

The child's solvability principle can be expressed as follows: Given sets S_x and S_y such that $C(S_x) < C(S_y)$, there exists a set S_a such that $C[A(S_x, S_a)] = C(S_y)$, and there exists a set S_m such that $S_m \subset S_y$ and $C[M(S_m, S_y)] = C(S_x)$.

The child's solvability principle might incorporate the concept of the inverse operation, that is, the concept that subtraction undoes the effect of addition and vice versa. If the child does in fact believe that $C(S_m) = C(S_a)$, then the concept of the inverse is implicit in the child's solvability principle. We have no real evidence one way or the other on this question, however. All we really know is that preschoolers believe that differences in numerosity can be eliminated by either removing something from the larger array or adding something to the smaller array. Whether or not the child believes that the numerosity of what must be removed is equivalent to the numerosity of what must be added is a question for further research.

It is worth commenting at this point on the difference between the notation we use to formalize our descriptions of the child's numerical reasoning principles and the conventional notation for expressing the laws of arithmetic. We believe that these notational differences reflect genuine differences in the conceptions being described. Foremost among these underlying differences is that the laws of arithmetic govern the handling of an abstraction called number whereas the child's numerical reasoning principles govern the handling of representations of numerosities. We believe that the child does not conceive of number except as a property of numerosities, that is, sets. The mathematician, for whom number has a reality that transcends numerosity, can think of adding and subtracting as abstract acts, without reference to numerosity. The preschooler, on the other hand, conceives only of adding and subtracting actual numerosities, that is, sets of *things*. Hence, in our notation for the child, the entities that operations and principles apply to are sets, with their numerosities indicated by subscripts.

One manifestation of the way the preschooler's numerical reasoning is tied to numerosity is the restrictions that appear in our notation for the preschooler's rule governing subtraction. The numerosity removed (subtracted) must be a proper subset of the numerosity from which it is removed. In other words, the first set within the parentheses in the expression M(,) must be a proper subset of the second. The dictum that subtraction may always be performed (even when b is greater than a) rules out such a restriction in modern formal arithmetic. As we noted, however, a remnant of such a restriction still influenced the thinking of some distinguished mathematicians in the early nineteenth century.

Associativity, Commutativity, and the Order Irrelevance Principle

The child believes that addition involves combining two disjoint nonempty sets. The uniting of sets is, quite independently of what the child thinks about it, a commutative procedure: The set formed by the union is the same regardless of which set is united with which. It is also an associative procedure: In forming the union of three sets, the result is the same regardless of which two are united first. Therefore, the additions performed by the child, insofar as he accurately evaluates the numerosity of the resulting unions, are associative and commutative. Whether the child recognizes this fact and makes use of it as a principle in his reasoning is another question. Like the question of whether the preschooler uses transitive inference (see Chapter 10), this question is moot. Because the child adds only specific (nonalgebraic) numerosities, it is not clear how to distinguish between commutativity and associativity as principles of reasoning and commutativity and associativity as inherent properties of the set-uniting procedure.

However, what the child believes about counting in conjunction with what he believes about addition produces something akin to the principles of commutativity and associativity. At least by the age of 4 or 5, children believe that the order in which you count the elements of a set does not affect the representation of numerosity you end up with. We have called this belief the order-irrelevance principle. Addition, in the child's view, involves uniting disjoint sets and then counting the elements of the resulting set. According to the order-irrelevance principle it does not matter whether in counting the union you first count the elements of one set and then the elements of the other or vice versa. Indeed, it does not matter if you skip around, counting first an element from one set, then an element from the other, then another element from the first, and so on. Thus, the child's order-irrelevance principle, when considered in the light of his addition procedure, is closely analogous to a belief in the commutativity of addition. Extending the same line of thought to the case in which the child adds three sets leads to the conclusion that the associativity of addition is also implicit in the child's numerical reasoning. The child believes, at least implicitly, that $C[A(S_x,S_y)] = C[A(S_y,S_x)]$ and that $C\{A[A(S_x,S_y),S_z]\} = C\{A[S_x,A(S_y,S_z)]\}$.

Closure

As noted earlier, the system established by the laws of arithmetic is, among other things, a group. It has all of the essential characteristics of a group, including *closure:* The result of adding any two numbers is always another number, that is another element of the set. Addition (and therefore subtraction) never opens the set; it never breaks out of the confines of the set of numbers by yielding an element that is not a number.

We know that the child's system of numerical reasoning is not closed with respect to subtraction. The constraint that the subtrahend must be a proper subset of the minuend explicitly violates the rule of closure. This lack of closure does not derive solely from the physical impossibility of removing more than x items from an x-item set. It is perfectly possible to remove x items from an x-item set, yet in the child's system this operation, when applied to two numerosities, yields a nonnumerosity, namely $\emptyset$, the empty set. In order to stay within the realm of numerosity, the child restricts subtraction to the removal of a proper subset, a subset that is not coextensive with the set itself.

The property of closure with respect to addition does not in and of itself require that there be an infinite or unending number of elements (numbers). The 10-step circular set we used to illustrate the concept of a group had only 10 elements (positions), yet it was closed under addition. In conjunction, however, the property of closure and the monotonic law require that the set of numbers be infinite. There cannot be a largest number. No matter how large a number is, it is always possible to add some other (nonzero) number to it (addition can always be performed), and the result will (by the monotonic law) necessarily be an even larger number.

We do not yet have strong evidence regarding whether the child's numerical reasoning forms a closed system with respect to addition. Diane Evans has been investigating this question. Her preliminary data indicate that most preschoolers do not recognize the infinite nature of number. Most preschoolers appear to think that there is a largest number. Many even think that they know what the largest number is. Such beliefs may reflect the child's inability to conceive of abstract numbers. As we will explain in the next chapter, we believe that the recognition of the necessarily unending nature of number is linked to the emergence of the ability to treat numbers algebraically rather than only as representatives of specific numerosities.

Formal Principles Linking Numbers to Numerosities

Formal arithmetic as such is a symbolic game. Certain symbols—whose meaning, denotation, or reference need not be considered at all—are manipulated according to certain rules—the laws of arithmetic. These symbols need not be thought of as representing numerosities. Indeed, these symbols may very well, in a particular instance, refer to irrational numbers, such as $\sqrt{2}$, which cannot represent numerosities in any ordinary sense of the term. There is no set whose numerosity is represented by the number $\sqrt{2}$. Nor can $\sqrt{2}$ represent some fraction of the elements in a set. The number 11/83, on the other hand, does represent a fraction of the elements in any set whose numerosity is represented by any integer multiple of 83, and $-2{,}000$ can represent the numerosity of the dollars in a badly overdrawn checking account. Thus, all of the rational numbers (positive integers, negative integers, fractions, and zero) can be thought of as representing numerosities. This is not true of the irrational numbers; yet the laws of arithmetic apply with equal facility and appropriateness whether the symbols being manipulated are rational or irrational numbers.

If modern man used numbers only to represent numerosity, his number system would not include irrational numbers. Modern man, however, also uses numbers to represent magnitudes, such as the length of a line segment, the speed of a car, and the strength of a material. This practice thas forced him to invent the irrational numbers. Without the irrational numbers, some lengths could not be represented numerically, some speeds could not be expressed numerically, and so on. Consider, for example, a square with a line drawn diagonally from one vertex to another. Let the length of each side of the square be represented by the number 1. Then, by the Pythagorean theorem, we know that the number that represents the length of the diagonal must have the property that when multiplied times itself it is equal to the sum of the squares of two of the sides: $1^2 + 1^2 = 2$. But, it can be proven very simply that none of the numbers that represent numerosities or fractions of numerosities—none of the rational numbers—has the requisite property.

The fact that the numbers we use to represent numerosities are not adequate to the job of representing lengths of lines, speeds, and so forth is a serious impediment to the application of numerical reasoning to aspects of the world other than numerosity. Yet civilized man, who surveys the land, who trades in copper and other continuous quantities,who has hit upon the idea of representing magnitudes numerically through measurement, repeatedly attempts to apply numerical reasoning principles to numbers that do not represent numerosities. When he does so, he stumbles on unanticipated perplexities. Consider the following facetious example: A man arrested while driving at a speed equal to $\sqrt{10{,}001}$ miles per hour in a

25-miles-per-hour zone might try to persuade the judge that the number that represented his speed could not be arrived at by any procedure that depended on counting and that therefore there was no specified procedure for deciding whether this number was greater than or less than 25. The chances of such a defense getting him off are remote, but the example suggests the nature of the difficulties posed by confining our use of the term *number* to representations of numerosities. The inability to represent numerically such measures as the length of the diagonal of the unit square stands as a major obstacle to the unification of two branches of mathematics—arithmetic and geometry. Yet Descartes's analytic geometry cries out for the construction of such a union and the calculus is unthinkable without it. In order to effect this union it is necessary to introduce numbers that do not represent either numerosities or fractions of numerosities, namely, the irrational numbers.

The irrational numbers, in other words, arose in connection with attempts to interpret the elements in arithmetic as representations of properties other than numerosity. And this fact serves to emphasize the more general point that if we want to render the game of arithmetic useful, then we are always going to have to *interpret* it. We are going to have to let the symbols in arithmetic refer to or stand for things that are outside the game, things like numerosities and lengths of lines. The process of interpreting the arithmetic system gives rise to considerations that do not arise so long as we focus only upon the game itself.

One such consideration is this: When are we justified in saying that two things are numerically equal? The laws of arithmetic have nothing to say about this question! Entities to be manipulated arithmetically either are or are not defined to be equal. If they are so defined, then the laws of arithmetic tell how they may be manipulated. These laws lay down no criteria for deciding whether entities outside the system (such as the numerosities of two sets) should or should not be regarded as numerically equal. That problem is a separate problem, one that arises only when we want to interpret the players (elements) in the arithmetic game as representative of something, in this case, numerosity.

The question of what criterion we should use to decide whether two real-world things (such as the numerosities of two real-world sets) are numerically equal is nonetheless profoundly important in any attempt to apply arithmetic to the world or to other domains of thought. Hence it cannot be ignored when we consider arithmetic as a system for reasoning about the world.

All of which brings us to a comparison of various methods for answering the question of when two numerosities are numerically equal. As we shall see, it is in regard to this question that the child's arithmetic reasoning departs most noticeably from formal arithmetic reasoning. The most common formal rule for deciding whether two sets have equal numerosity is this: Two sets have equal numerosity (are numerically equal) if the elements of one set can be placed in one-to-one correspondence with the elements of the other set. Note that the decision about numerical equivalence is reached in the absence of any representation of numerosity. Indeed, the representation of the numerosity of a set is achieved by establishing a numerical equivalence (a one-to-one correspondence) between that numerosity and one of a sequence of abstract reference sets. The sequence of abstract reference sets is constructed for the sole purpose of formally determining which numbers refer to which numerosities. This sequence of reference sets is frequently constructed according to the following procedure. The number 0 is defined as the representative of the empty set. The number 1 is defined as the representative of the set of all sets that can be placed in one-to-one correspondence with the set that contains the empty set as its only member. This curious set, it should be noted, is not the empty set. Unlike the empty set, it has a member, namely, the empty set. The number 2 is defined as the representative of all sets that can be placed in one-to-one correspondence with the set that contains as its only members the empty set and the set that contains the empty set. And so on and on, to ever more deeply embedded empty sets.

This standard procedure at first glance strikes most people as an extremely artificial way of relating numbers to the numerosities they are to represent. It is, however, not at all easy to suggest an alternative that does not have some hopelessly provincial feature, such as being tied to the count words of the proposer's language. The standard procedure has two advantages. First, it recognizes the fact that representations of numerosity presuppose the establishment of a numerical equivalence based on one-to-one correspondence. (As we will note later, this is true for the child's procedure as well as for the standard procedure. The child's counting procedure establishes a one-to-one correspondence between elements of the count sequence and the elements of the set being counted. Unlike the more reflective professional mathematician, however, the child appears to take no cognizance of this fact.)

The second advantage of one-to-one correspondence as a criterion of equivalent numerosity is that it can be applied to sets with infinite numerosity, for which one could never obtain a representation of numerosity by the counting procedure we have described. This second advantage is of much greater importance to the professional mathematician than the relatively picky first advantage. It obtains, however, only if we construe broadly the phrase "can be placed in one-to-one correspondence." The mathematician George Cantor has shown that if we interpret this phrase to mean that a rule can be specified assigning to each member of one set a unique corresponding member of the other set, then we can apply the criterion of one-to-one correspondence to the question of whether or not two infinite sets have equal numerosity. If, however, we take the phrase to mean that we can actually construct the complete one-to-one correspondence for all the world to behold, then the one-to-one correspondence criterion cannot be applied to infinite sets, and only the minor first advantage remains in its favor.

All of those who favor the broad construal of one-to-one correspondence, and this group includes most but not all living mathematicians, must accustom themselves to some startling consequences. One consequence is that the numerosity of the set of all rational numbers is *equal* to the numerosity of any of its infinite proper subsets. A proper subset of a set is first of all a subset; that is, all of its members are also members of the set. A proper subset has the further constraint that, by definition, it does not contain all of the members of the set. Thus, for example, the set of all perfect squares, 1, 4, 9, 16, 25, and so on, is an emphatically proper subset of the set of all rational numbers. Yet, under the broad construal of the one-to-one correspondence criterion, the numerosity of the set of all perfect squares is *equal* to the numerosity of the set of all positive integers. Setting each positive integer in correspondence with its own square establishes the necessary one-to-one correspondence. By this rule, one and only one perfect square corresponds to each positive integer and one and only one positive integer corresponds to each perfect square. Given any positive integer, the rule identifies a unique perfect square that corresponds to that integer; given any perfect square, the rule specifies a unique positive integer that corresponds to that perfect square. By devising a subtler rule, Cantor also proved that the numerosity of the set of perfect squares is equal to the numerosity of the set of all rational numbers. This despite the fact that, on the one hand, every perfect square is a rational number, while on the other hand, there are infinitely many rational numbers that are not perfect squares.

The advantage that makes most mathematicians willing to live with these unsettling consequences is that, as Cantor also proved, the set of all *real* numbers (the rationals plus the irrationals) can be placed in one-to-one correspondence with the points on a line. Thus numbers can represent all possible line segments. In modern terminology, the real numbers and the points on a line are both *uncountably* infinite. The proof that the set of all real numbers is numerous enough that every point on a line can be represented by a real number lays the groundwork for uniting geometry and arithmetic. It links the concept of number and the concept of linear magnitude. This union establishes a rigorous conceptual foundation for the calculus.

As one might suspect from the fact that no rational number can represent the length of the diagonal of the unit square, it is not possible to establish a one-to-one correspondence between the set of all rational numbers and the set of all points on a line. The numerosity of the set of all real numbers is greater than the numerosity of the set of all rational numbers. In modern terminology, the rational numbers constitute a mere countable infinity. A countable infinity is no match for an uncountable infinity, such as the points on a line. This usage of *countable* and uncountable highlights the difference between the narrow and the broad construals of one-to-one correspondence. The rational numbers are not countable in the usual sense: It is not possible to obtain the cardinal number that represents their numerosity. They are countable, however, in Cantor's sense: It is possible to specify a rule that assigns a unique count number (a unique positive integer) to each rational number.

Those who find Cantor's use of one-to-one correspondence bewildering, outrageous, or both may take comfort from the fact that a mathematician as distinguished as Poincaré thought that Cantor's theory would be regarded by later generations of mathematicians as "a disease from which one has recovered" (Kline, 1972, p. 1003). The broad construal of one-to-one correspondence and its consequences take some getting used to. This fact may be a reason for doubting that the untutored human mind regards one-to-one correspondence as the ultimate criterion for numerical equivalence. We hasten to add that whether the untutored or unreflective human mind does or does not regard one-to-one correspondence as the inner essence of numerical equivalence is of concern only to psychologists, not to mathematicians as mathematicians. In this book, we are concerned with numerical reasoning as a psychological phenomenon. We are not concerned with the proper foundations of numerical epistemology.

The point of this digression into the application of the one-to-one correspondence principle to the problem of deciding which infinite sets have equal numerosity is to show the central role played by one-to-one correspondence in mediating between the world and arithmetic principles. The centrality of one-to-one correspondence is not merely a matter of convention, an arbitrary choice from among many other equally satisfactory mediating principles. Only by adopting the broad construal of one-to-one correspondence as the foundation of numerical equivalence could turn-of-the-century mathematicians place the calculus upon a conceptually rigorous foundation.

The formal procedure for determining the order relation between two nonequivalent numerosities also rests on the use of one-to-one equivalences. A numerosity n is said to be less than a numerosity m if n can be placed in one-to-one correspondence with a subset of m but m cannot be placed in one-to-one correspondence with a subset of n. By this criterion, Cantor proved that the set of all rational numbers is less than the set of all real numbers. He also proved that there "exist" an infinite number of ever more numerous infinities, which in their infinite totality comprise the transfinite numbers. This proof shows how remote the modern mathematical concept of number is from the layman's conception. The modern conception was carried to these remote and airy abstractions on the powerful wings created by the broad construal of one-to-one correspondence.

The Child's Criteria for Numerical Equivalence and Numerical Order

The preschooler's normal principle for determining whether two sets are numerically equal is "Count them and see." Two sets have equal numerosity if when counted they both yield the same cardinal numeron. In the preschooler's method the decision regarding numerical equivalence is possible only after one has obtained representations of the numerosities. In the mathematician's procedure the reverse is true; the decision regarding numerical equivalence is prior to and does not require representations of numerosity.

As indicated, the child's procedure actually presupposes the establishment of a one-to-one correspondence. In counting, the child establishes a one-to-one correspondence between the elements in his count sequence and the elements in the set being counted. From a logical point of view, the child's procedure for deciding numerical equivalence depends on the fact that the numerosities of both sets can be placed in a relation of one-to-one equivalence with the same set of counting tags. But the child does not ordinarily take cognizance of the transitivity of one-to-one correspondence. He ignores or is indifferent to the fact that the cardinal numerons representing two equally numerous sets are identical precisely because both sets have been placed in one-to-one correspondence with a count sequence that terminates with that cardinal numeron.

The preschooler may under some circumstances be persuaded that two sets are numerically equal solely because he has observed that they may be placed in one-to-one correspondence. In a sense, however, as we suggest in the next chapter, the child does not know what to make of this judgment. Despite his recognition of the numerical equivalence, he lacks representations of the numerosities; and the preschooler's numerical reasoning principles, at least in their initial stage, are geared to operate only on representations of specific numerosities, that is, only on cardinal numerons.

The child's principle for determining the ordering of two nonequivalent numerosities is that the numerosity represented by the earlier numeron is less than the numerosity represented by the later numeron, where earlier and later refer to position in the count sequence. Again, the mathematician's use of one-to-one correspondence in this connection is implicit in the child's procedure. The cardinal numeron of the smaller set is earlier than that of the larger set, because, in the act of counting the two sets, the preschooler has placed the smaller set in one-to-one correspondence with a subset of the set of counting tags required to achieve a one-to-one correspondence with the larger set. Again, however, the preschooler is not a mathematical logician; he takes no cognizance of the logical priority of the ordering criterion based on one-to-one correspondence.

Overview

We have insisted on the importance of following the lead of modern mathematics in distinguishing sharply between the principles of numerical reasoning, on the one hand, and the principles by which representations of numerosity are obtained, on the other hand. It is nonetheless obvious that the two kinds of principles interact strongly in determining the preschooler's conceptual scheme for dealing with number.

The young child's principles of numerical reasoning are designed to be applied to representations of numerosity obtained by counting. Counting yields only what mathematicians term the natural numbers, namely, the positive integers. Therefore, in place of the single law of addition recognized by modern arithmetic, the preschooler has three operations, addition, subtraction, and solving. Counting does not yield negative numbers; hence the child does not conceive of subtraction as adding a negative number. His principles could hardly be formulated so as to require elements that his counting principles cannot supply. Instead, the child conceives of subtraction as an entirely separate operation; furthermore, an operation that can be performed only when the subtrahend is a proper subset of the minuend. Similarly, the child cannot conceive of solving for the difference between a and b in terms of finding the number that when added to a makes a sum equal to b. If $a > b$, the number that must be added is negative. Rather, the child conceives of either finding a numerosity that can be subtracted from the larger numerosity to yield the lesser numerosity or finding a numerosity that can be added to the lesser to yield the larger.

The addition operation in arithmetic also subsumes the identity operation, because in arithmetic one has the number zero, the so-called identity element. However, counting does not yield a numeron or numerlog that is the equivalent of zero. Hence, the description of the child's numerical reasoning requires a fourth class of operations, identity operations.

Another manifestation of the impact of number-abstracting principles on numerical reasoning principles is seen in the domain of the relations. It is perhaps impossible to determine whether the young child recognizes the transitivity and symmetry of the numerical equivalence relations, because his reasoning principles come into play only with representations of specific numerosities. One cannot even state the symmetry of numerical equivalence using specific numerosities. The statement "if $3 = 3$, then $3 = 3$" is nonsensical; whereas the statement "if $a = b$, then $b = a$" is not. It is similarly nonsensical to try to state the transitivity of a numerical relation using elements that represent specific numerosities. What is one to make of the statement "if $3 = 3$ and $3 = 3$, then $3 = 3$"? The statement of the transitivity of the order relation, "if $3 < 4$ and $4 < 5$, then $3 < 5$" does not seem nonsensical, but how can one prevent the child from observing that $3 < 5$ quite independently of the relations between 3 and 4 and 4 and 5?

The final, and perhaps the developmentally most interesting, consequence of the young child's need for representations of specific numerosities is that it prevents the child from bringing his numerical reasoning principles to bear upon judgments of numerical equivalence that derive solely from the perception of one-to-one correspondence. The perception of one-to-one correspondence may yield a judgment of numerical equivalence, but the numerical reasoning principles utilize only representations of numerosity, and the observation of one-to-one equivalence does not yield a representation of numerosity. One may watch a large troop of cavalry go by and know that there is one horse for every rider. Such knowledge does not depend on the specific number of horses or riders. Yet the specific numerosity is precisely what the preschooler seems to require before he can reason numerically. Judgments of numerical equivalence based on one-to-one correspondence do not have access to the computational routines that carry out the preschooler's numerical reasoning. We argue in Chapter 12 that these judgments begin to gain access to the computational machinery only as the child approaches school age. We suggest that passing Piaget's conservation test marks this gaining of access. We join with Piaget in interpreting this change as an important developmental milestone, although we disagree with Piaget about the road by which the child reaches this milestone.

Chapter Twelve

What Develops and How

We began this book with an exhortation to researchers in the field of cognitive development to look at preschoolers in their own right and not just as a group who fail tasks that older children pass. We suggested that failure on a particular task did not necessarily indicate a complete absence of the cognitive capacity supposedly tapped by that task. It seemed reasonable that if tasks were designed with the preschooler in mind they might well show that young children possess some of the capacity they have been presumed to lack. Throughout the book the reader will find young children who know much about counting; who can reason about number; and who hardly lack a concept of number. Our research was premised on the assumption that once the data began to come in we would be in a better position to determine the *what* and *how* of numerical development. Our idea was that it should be easier to provide such accounts given a description of some ability, rather than one of no ability. Not only would we be able to say that young children could do such and such; we would also be able to examine their accomplishments for clues as to what might develop. We might also be better able to determine why young children do so poorly on the standard tasks that have been used. Having found tasks that shared some features with the standard ones, we could chisel away at the theoretical sources of difficulty, dropping some possibilities while keeping others and thereby constraining the number of theoretical accounts of what develops. In short, we anticipated using evidence on the numerical abilities of preschoolers to make progress toward accounting for the nature and development of numerical abilities in yet older children.

Counting and Other Input-Output Skills

The How-to-Count Principles: Principle versus Skill

Children as young as 2½ years use the how-to-count principles. When confronted with small set sizes, they tend to use as many tags as there are objects to tag; they tend to assign unique tags; they tend to use a stably ordered list of tags; and they often indicate that the last tag assigned in a given enumeration represents the cardinal number of the set of objects. This is not to say that they count as adults do; nor is it to say that they never err.

Actually, it is fortunate that the very young child's counts contain errors, sometimes to such an extent that they do not at first blush resemble proper counts. The kinds of errors that appear in the application of the counting principles and the seemingly idiosyncratic behaviors give us insight into the nature of what develops. We begin by focusing on one feature of many young children's counting, the use of idiosyncratic lists. Recall that some children counted a two-item array by saying "two, six," or "*A, B*," and a three-item array by saying "two, six, ten," or "*A, B, C*." Upon inspecting such count trials we discovered a remarkable fact: These children used their idiosyncratic lists in a systematic fashion. The list was stably ordered, that is, the same sequence of tags occurred trial after trial. What is more, the child assigned as many tags from his list as was necessary in a given situation. Thus if the child saw a two-item array, he used the first two tags from his list; if he saw a three-item array, he used the first three tags from his list; and so on. Finally, when asked "how many?" he responded by repeating the last tag used. It was not uncommon for the experimenter to repeat the question after the child said "six" or "*C*" in the presence of a two-item array. And the child obliged by repeating what he had just said.

Why do we focus on such idiosyncratic responses? Because they are important evidence for our thesis that children as young as 2½ apply the how-to-count principles and that much of the development around these principles involves skill at applying them. Why do we grant such young children knowledge of these principles? And how does skill in their use come about? First, the reasons for granting the principles.

A brief review of the how-to-count principles helps make clear our reasons for concluding that young children make use of them. First, there is the one-one principle, which involves the assignment of unique tags to each and every item of the array. Second, there is the stable-order principle, which requires that the tags be assigned in the same order across counts. Finally, there is the cardinal principle, which involves singling out the last tag assigned on a given count to represent the cardinal number of the array in question. When we analyze count trials for evidence about these principles, we find two sorts of children—especially in the youngest age group. One sort of child counts in the conventional way. A two-item array yields a "one, two; two" answer; a three-item array yields a "one, two, three; three" answer. We cannot and do not rest our case on such evidence. One could account for such data by simply assuming that the child who counts "one, two" or "one, two, three" has committed these words to rote memory. One could assume that the memorizing of the conventional list in the conventional order precedes the induction of counting principles. In other words, one could argue that skill in reciting count-word sequences precedes and forms a basis for the induction of counting principles. We, however, advance the opposite thesis: A knowledge of counting principles forms the basis for the acquisition of counting skill.

It is the second kind of child who provides the critical data for postulating the availability of a counting scheme that directs the learning of the conventional list of count words. These are the children who use idiosyncratic lists. Some of these children produce lists that they are unlikely ever to have heard used for enumeration. An occasional child uses the alphabet, which is a series that they indeed have heard but not in the context of counting. The significant fact about these lists is that they are used in a way that is prescribed by the counting principles. It seems reasonable to conclude that the availability of the principles governs such behavior. Any other conclusion would require postulating the existence of systematic behavior that resembles counting and occurs by chance. Granted, when we first encountered such behaviors we thought them random uses of number words and the alphabet. But when we subjected them to analyses suggested by the counting principles, we discovered that such children were telling us, in their own way, what they knew about counting.

Analysis of the types of counting errors made provides further evidence that young children can and do apply principles when counting. Consider first the errors in applying the one-one principle. In the main they are of two kinds: partitioning errors and coordination errors. The locus and nature of these errors are quite restricted. Partitioning errors occur as the child tags successive adjacent items. As he moves from item to item he sometimes double-counts an item or skips an item. He hardly ever skips around, going back and forth in the array. Together these facts lead us to conclude that the child is guided by a partitioning rule; if not, he should skip around a great deal and should show little evidence of systematically proceeding from item to item. The double count or omit could occur because the child momentarily loses track of where he is or because he is sloppy in his transfer of items from the to-be-counted to the already-counted sets. What might make him sloppy? We think there are a variety of possibilities. He might hesitate to move each item aside as it is counted. He might point too rapidly. In both of these cases, his error might be thought of as resulting from poorly executed strategies for accomplishing the partitioning required by the one-one principle. But the fact remains that in general the child does move systematically from item to item and point as he does so. Why would he behave this way if he were not following the one-one principle? To be sure, there is room for becoming more skillful in the execution. Indeed, this need for greater skill seems to be a major target for development.

The coordination errors can be thought of in a similar way. To show why, it helps to note what kinds of errors do *not* occur. Children seldom produce totally asynchronous counts; in general they point to and touch a single item (the partitioning component) *and* state a single numerlog (the tagging component). Difficulty occurs to some extent at the starting of a count and to a larger extent at the stopping of a count. A child may start to point before he starts to recite tags, or vice versa; and if he does he may take some time to get beyond the first item. But then things proceed smoothly. Trouble reappears only when he nears the end of a count. He may run over or under by one item—but not by two, three, four, or more items. We take this behavior as an indication that the child realizes that the tagging and partitioning processes should stop together. And they almost do. The lack of coordination is limited to the final item. We think this type of error reflects a faulty stop rule on the motor side—a conjecture that is consistent with Russian work on preschoolers (for example, Luria, 1961). Again we see a need for the child to develop skill in applying a principle that already guides his behavior.

Since the child strives for a one-one tagging performance before he is skillful at achieving such a performance, we conclude that the one-one principle guides the acquisition of the skill. What about the stable-order and cardinal principles? Does it make sense to argue that these too are available to the young child and that they guide the quest for skill in their application?

When we introduced the stable-order principle we made much of the fact that it was possible to apply it without using conventional count words in the conventional order. In our observations of how young children count, we found them using elements from two stably ordered lists, the count words and the alphabet. They did not necessarily use these elements as adults do, however. When counting the number of items in a set, adults do not use the letters of the alphabet as tags. To be sure, under some conditions we do use the alphabet to keep track of cases that are related. We identify sections of written material with *A, B, C,* and so on. Likewise we use the letters of the alphabet to rank performances; witness their use as grades. But we do not answer questions about numerosity, such as "how many apples do you need?" by stating a letter of the alphabet (though it is perhaps worth noting that the Greeks did). So the young child who does so is not acting in a conventional way. Why might he use the alphabet in this way? The conclusion seems inescapable that his behavior is governed by a principle that specifies the use of a stably ordered list but leaves open which stably ordered list should be used. The principle is neutral with respect to type of tag; it simply requires that the tags used be drawn from a stably ordered list. The alphabet is such a list, and it is one list that young children in our culture encounter at an early age. To explain its use in counting, we postulate the availability of a stable-order principle that guides the assimilation of relevant environmental inputs. Without such an assumption, the fact that some very young children use the alphabet to tag items in counting remains unexplained.

It is not just the occasional child who uses the alphabet when counting who serves as evidence for our thesis. So do children who use the conventional count words in an unconventional, idiosyncratic order. To make sense of very young children's tendency to work with idiosyncratic but stably ordered lists of number words, we assume that the behavior is guided by the stable-order principle. This assumption is consistent with another result. In our report of the 2-year-olds' performance in the videotape experiment, we noted that children who used idiosyncratic lists were better able to apply the stable-order principle than were children who used conventional lists of number words. A child who used an idiosyncratic string used them in the same order trial after trial. In contrast, the child who used the conventional string tended to be shaky in following the stable-order rule; he would use the conventional list often enough for us to discern a tendency to do so, but during many count trials he slipped up, sometimes skipping a tag in the list, sometimes using the tags in a seemingly random sequence, and so on.

Why is the user of an idiosyncratic list more consistent in his application of the stable-order principle? Adults who are allowed to organize material in their own way learn it more readily than adults who are required to detect and apply an experimenter's organization of the material (Mandler and Pearlstone, 1966). The imposition of an external organization interferes with an already available one. If we grant that the young child's counting is governed very early by a stable-order principle, then the tendency to do better with idiosyncratic lists is explained. The child remembers a list of his own making better than one imposed from outside. But it will not do to stay with an idiosyncratic list. Such lists are consistent with the requirements of the stable-order principle, but they are *not* consistent with the requirements of communication. Imagine what it would be like to do arithmetic if different individuals used different lists of number words. The young child has to switch to using a conventional list. Otherwise, his knowledge of the counting procedure is likely to go unrecognized by those around him. He probably will require considerable practice before he is skilled in his use of the conventional list, but he probably would need far more practice if his learning were not guided by the stable-order principle.

The case for assuming that the cardinal principle is used by very young children is weaker. Some young children do indeed repeat the last tag in their list, be the list conventional or idiosyncratic. But the tendency of young children to apply the cardinal principle is limited. The smaller the set size, the more likely is the child to apply the cardinal principle. We think that the reason for this fact can be traced in part to the way the how-to-count principles are interrelated. In order to apply the cardinal principle the child must likewise apply the one-one and stable-order principles. Having successfully tagged each item in the set, he must then repeat the last tag used. Accordingly, the skilled use of the cardinal principle should be dependent on the skilled use of the other two how-to-count principles.

The extent to which a child honored the cardinal principle did indeed depend on his skill at using the other two principles. Children who erred in applying the one-one or stable-order principle were unlikely to apply the cardinal principle. Thus it makes sense that we see more evidence for the application of the cardinal principle when the sets to be counted are small. For the smaller set sizes, the children have already developed the skill to apply the one-one and stable-order principles in concert.

The How-to-Count Principles Form a Scheme

Whence comes the skill at applying the how-to-count principles? What can we conclude from our observation that very young children use the counting principles but apply them none too skillfully? We believe that our results support the view that these principles constitute a scheme in the Piagetian sense. The principles guide and structure the child's counting behavior, serve as a reference against which the child can evaluate this actual counting behavior, and motivate that behavior.

We have already highlighted the role of the counting scheme in structuring the child's observable behavior. That the scheme is also used as a reference or touchstone is evidenced by the child's frequent self-corrections. The motivating function of the scheme is evidenced by the frequency of spontaneous and seemingly purposeless counting by young children—of the number of toys they have, the number of steps in the house, the number of candies, cracks in the sidewalk, and so on. In short, we assume that the how-to-count principles constitute a scheme in the Piagetian sense because they reveal the characteristic properties that Piaget attributes to schemes. For Piaget children practice schemes; for us the counting principles lead to spontaneous counting. For Piaget schemes cause children to search out and assimilate aspects of the environment that fit the schemes; for us the counting scheme leads children to use the alphabet or idiosyncratic count lists. For Piaget the environment forces accommodation; for us the environment forces the child to adopt a conventional counting sequence. For Piaget the scheme serves as a structure that is in a state of equilibrium or disequilibrium; for us, the counting scheme serves as a source of reference against which the child can evaluate and refine his counting behavior.

Perhaps we force the case when we draw such parallels. The important thing is to recognize that young children are motivated to count on their own, that they self-correct, and that they eventually come to count accurately whether or not they are in nursery school and whether or not they watch "Sesame Street." And we think something akin to the Piagetian notions of scheme, assimilation, and accommodation is involved. By postulating such general ideas we are able to make sense of what we observe and to understand how children might acquire skill at counting.

Consider the assumption that the counting scheme motivates counting behavior in light of children's spontaneous attempts to count. Such spontaneous counting seems purposeless, but developmentally it may be far from purposeless. It provides the children with practice. Add to this behavior the tendency to self-correct—again presumably because the counting scheme provides a point of reference—and what do we have? A procedure for producing practice trials, many of which can be corrected by the child himself. This is precisely what we want. For we are investigating how young children become more skillful at counting. They practice—apparently because they "want to." They self-correct. Their scheme plays the role of a personal tutor who both goads and guides. All of this, plus an environment that contains stably ordered lists to be assimilated, enables children to become skillful in the application of the how-to-count scheme.

Our assertion that early counting behaviors and practice at them are driven by underlying principles or a scheme may be novel in the context of discussions on counting. The outlines of the argument, however, are not all that novel. We have already shown that they resemble Piagetian notions. It can also be said that our views are much like those of Noam Chomsky (1965) on language acquisition, which derive support from a set of observations similar to ours. For example, young children say things that they have never heard but that are rule-governed ("I runned"); and they self-correct and rehearse without any obvious stimulus from the environment (Weir, 1962). Somehow there has been an unstated assumption that in preschoolers these kinds of behavior are language specific. In fact such behavior—especially the endless tendency to practice—is probably characteristic of a host of cognitive activities in young children.

Development of the How-to-Count Principles

We believe that the young child's spontaneous counting is of considerable import for the development of counting. The tendency to count without any apparent motive provides the child with practice in applying the counting procedure. Coaches and music teachers are unanimous in their testimony that, when it comes to the perfection of motor skills and their coordination, there is no substitute for practice. We have often noted the extensive use of motor processes on the part of young children who count. They point when they start to recite tags, they point as they tag and partition items in an array, and they stop pointing when they run out of items to tag. The use of pointing and the tendency to touch or even move items can be thought of as motor strategies that help the child coordinate the partitioning and tagging processes involved in the application of the one-one principle. Likewise they probably aid in the execution of the component processes themselves. The child's pointing is therefore helpful, perhaps even necessary at first. Pointing externalizes the coordination problem that is inherent in proper counting. The points must be coordinated with the verbal output, which in turn must be coordinated with the partitioning process. Many of the counting errors of younger children stem from failing to start and stop the tag-withdrawal process simultaneously with the partitioning process and consequently with the use of points or related hand movements. How might the child learn to avoid such errors? The errors stem in part from poor motor coordination, which may presumably be improved by practice. And from within the child comes the necessary impetus to practice. Thus the child provides himself with the very kind of experience he needs. There is no substitute for practice in learning to coordinate component skills or to do precisely what has been done imprecisely. Since these are the very types of learning that contribute to skill in applying the one-one principle, the development of counting skill should be traceable to the opportunity for and the extent of practice.

Besides motivating and correcting the practice through which the child refines the motor components of his counting skills, the counting scheme motivates the child's search for and assimilation of stably ordered lists for use as tags. Developmentally, one can follow the vicissitudes of this search between the ages of 2 and 5 and then beyond. The counting behavior of the very youngest children incorporates words from certain stable lists in the verbal environment, most noticeably the count words and the names for the letters of the alphabet. In our experience, the use of the alphabet in counting occurs only in the younger children. It seems very unlikely that adults use the alphabet for enumeration when speaking to young children. None but the most academic (and pompous) of parents is likely to say to his young child, "I want you to (*a*) wash your hands and (*b*) eat your applesauce." The use of letter names in counting emphasizes to us the validity of regarding the young child as a being whose learning of stable verbal count sequences is guided by a stable-order principle in search of a stably ordered list.

Almost all children settle at a fairly early age on the count words themselves, thereby narrowing their search for a list to the culturally acceptable ball park. Their problem then becomes learning the traditional words in the traditional sequence. Many young children pass through a stage in which they employ stable but idiosyncratic orderings. Such behavior again supports the view that the child's learning is guided by his counting scheme. The scheme specifies the structure of what is to be learned. In the case under discussion it specifies the learning of a stably ordered list. The child who imposes idiosyncratic orderings on the count words is following the dictates of his scheme. However, the idiosyncracy of his ordering is likely to interfere with his assimilation of extensions of the list. If the child's ordering runs "one, two, three, eight, five," and he attempts to extend his list to eight tags by assimilating further count words in the order in which he encounters them in the environment, he must run into conflict when he encounters the word *eight* used in eighth position. If he adds this word to the eighth position in his own list, which already contains the same word in the fourth position, he violates the principle that the tags must be distinct. This sort of conflict, together with corrections from parents, teachers, and older children, should lead the child to accommodate the order of his own list to the order of the traditional list.

As we have emphasized, the counting scheme motivates the learning of and practice with an environmentally given count sequence. However, this learning cannot proceed very far before it is stymied by an inherent limitation of human memory. Human memory cannot readily store stable, recallable lists of items when there is no generative rule underlying the ordering of the list. The tension between the counting scheme's requirements and the limitations of human memory has led to the universal adoption of rules for generating words to fill the higher positions in the count sequence. The generative rules, however, do require the initial learning of a base sequence of numerlogs. The length of the base sequence of numerlogs varies from one linguistic environment to the next. It may be as small as 2 in some cultures and as large as 20 in others (Zaslavsky, 1973). Most commonly the sequence that must be learned is 5 to 15 items long. The sequence often corresponds approximately, but not exactly, to what mathematicians term the base of a number system. The English linguistic community uses a base 10 system. However, historical changes in the pronunciation of the first few count words beyond *ten* have obscured their derivation from a generative rule. Very few modern English speakers perceive eleven and twelve as contractions on one-left and two-left (that is, left after ten). Thus these count words should probably be considered among those that are learned by rote. Fourteen is the first English count word whose derivation from an earlier member of the rote sequence is completely transparent. It is preceded and followed by somewhat less obvious derivations, thirteen and fifteen. Only with sixteen does the generative rule emerge in an unambiguous and regularly repeated fashion. Its emergence, however, does not mark the limits of the rote learning required by the count sequence. The child must still learn the new words that represent the successive exhaustions of the base sequence: twenty, thirty, and so on. Of course, these words have their own underlying rule-based derivation. They may be mastered through a perception of the derivation after exposure to the first few instances: twenty, thirty, forty. Once this rule is surmised the child need only learn the words for the second, third, sixth, and so on powers of ten. All other count words can be derived from the application of the generative rules embodied in the already mastered count words.

Thus the counting scheme, presumably together with the development of the child's ability to perceive underlying generative rules, leads to the mastery of a count sequence that requires a limited amount of rote learning but is capable of being extended almost indefinitely. The rote part of this learning proceeds slowly and must surely be aided by the young child's tendency to practice and self-correct. Extensions of the count sequence that depend upon a grasping of the generative rule, such as the extension from fourteen to sixteen on up to nineteen, probably occur rapidly once the relevant generative rule is grasped. At twenty there might very well be a pause. Once the child masters twenty, twenty-one, and perhaps twenty-two, the sequence extends rapidly again to twenty-nine.

Up to this point we have focused on developmental processes that improve the components of the counting scheme and the coordination of those components. These processes involve learning the system of counting tags, perfecting the use of the individual principles that constitute counting, and tightening the coordination of the ensemble. The result of this line of development is to make the counting procedure an efficient and reliable routine.

This routinization of the counting procedure is manifested in several developmental trends. One early manifestation is an increase in the application of the cardinal rule. Very young children indicate that the cardinal principle is part of their counting scheme by counting aloud and repeating or stressing the final numerlog in the sequence of count words. They sometimes seem to forget to apply the cardinal rule, however, presumably because they are absorbed in carrying out the parts of the counting procedure that must precede it. As the other aspects of counting become more routinized, the cardinal rule is applied with greater regularity. As this development proceeds, a related trend becomes noticeable: The child's tendency to count aloud diminishes. We assume that the strong tendency of very young children to count aloud, even when counting the smallest sets, is an index of the difficulty that the counting procedure presents for the young child and the consequent demands it makes upon the young child's attention. With practice, the procedure becomes more routinized and thus requires less meticulous attention. The child is then free to focus more and more on the end result of the counting procedure—the cardinal numeron that represents the numerosity of the set. In the transition from counting aloud and then stating the cardinal numeron to stating the cardinal numeron without having counted aloud, the child passes through an intermediate stage in which he counts out loud when he first encounters a set but merely states the cardinal number on subsequent encounters. The increased tendency to state the cardinal number without first counting aloud probably reflects the redistribution of attention that is a consequence of the increased routinization of the counting procedure. (See LaBerge and Samuels, 1974, for a comparable argument regarding the development of reading abilities.)

The Abstraction Principle

The abstraction principle deals with the question of when the counting procedure can be used—of what constitutes a collection of countables. We have called it the what-to-count principle. It is common for developmentalists to assume that, for the young child, the category of things that can be counted is extremely restricted. It has been assumed that initially the young child will only count collections of identical objects. Thus, even differences in color, as in an array of red, green, and yellow apples, might stand in the way of the young child counting all the apples. He might count the red ones, the green ones, and then the yellow ones, failing to see that he could simply count apples or *things*.

We have sought to determine whether in fact young children, as a matter of principle, place severe restrictions on the nature of objects that can be counted together. We have compared the ability of young children to work with heterogeneous sets versus homogeneous sets.

Over and over again we find that young children can and do abstract numerical representations of heterogeneous sets. Variations in color or item type have little, if any, effect on the accuracy of a count or the ability to state the number of items contained in a display—as long as we take care to avoid procedures that might lead the child to think that we expect him to restrict a count to like items. Heterogeneity appears to have an adverse effect only if experimental conditions induce the child to focus on homogeneous items. For example, Gast (1957) tested children first with homogeneous arrays and second with heterogeneous arrays.

Although we are convinced that preschoolers can be quite flexible about what they will collect together for counting purposes, we do not wish to give the impression that they realize that almost any imaginable set of items (physical and nonphysical) can in principle be counted. In the absence of data on the restrictions they may very well place on the domain of countables, we can only make guesses about what these might be.

In our research we have used two types of heterogeneous arrays. In some cases the objects varied in color: green and red mice or red, yellow, blue, and white chips. In other cases the objects varied in both item type and color. Classifying such types of objects together is relatively simple. Where color varies, the child can count mice or circles (chips). Where item type varies, the child can count things or toys. It is not difficult to come up with a term that applies to all the objects shown. But we can think of heterogeneous collections that might not lend themselves to obvious labels. Consider the case of imaginary playmates. Children might resist the idea that these are things if they believe that nonphysical entities cannot be things. Such an outcome would be perfectly consistent with many theories that assume a developmental course from concrete to abstract thinking. Thus we might expect to see children restricting their use of the counting procedure to things in the world.

What other restrictions might young children apply? Perhaps they separate some classes of things from other classes of things, say trees from people. Or perhaps they insist that people are not things (compare Keil, 1977).

We don't know whether young children do apply such restrictions. We simply want to suggest types of restrictions that might be applied as opposed to those that, as we have shown, are not applied. Variations in color, size, type of toy, are not obstacles to counting because they do not prevent a child from assigning a common generic term to all items in a collection. But there is more to the account of why certain heterogeneous items can be counted together. Trees and flowers and chairs could very well constitute a tolerable collection for a child, as could all the objects in the child's bedroom. All of these objects can be called *things.* We are suggesting that children who say that certain objects are not things will resist including such objects in a count along with objects that they readily call things. Thus the boundary conditions for restrictions can be defined in terms of (*a*) an inability to assign a common generic term to a set of different objects or (*b*) an inability to assign the term *things* to all of the objects in a given collection. In this context it no longer makes sense to talk of restrictions in reference to attributes such as color and size. And our failure to find children who refused to count arrays of objects that varied along such dimensions is explained.

An adult might think of yet another kind of restriction, as Figure 12.1 illustrates.

Figure 12.1 Drawing by Dana Fradon; © 1976 The New Yorker Magazine, Inc.

As adults we know full well that we cannot add apples and oranges. The operation of addition can only be applied to like types. How might such a restriction rule bear on our discussion of the abstraction principle? Recall that young children use a counting algorithm when adding. So perhaps when they add they too refuse to add apples and oranges. We think not. Indeed we can now offer an explanation of young children's puzzlement when they are told they cannot add apples and oranges. They have always counted the apples, the oranges, the bananas, and so on as things in Mommy's refrigerator. So what can the teacher be talking about? Their use of the abstraction principle is still implicit. When counting the apples and oranges they count things or pieces of fruit. They assign a common hierarchical term to the members of the collection even though they may not realize it. In Rozin's terms (1976), they have not yet *accessed* a principle that governs their behavior. With development they will, and therefore with development will come an explicit understanding of what it means to say "apples and oranges don't mix."

The Order-Irrelevance Principle

We assume that the individual who understands counting as a procedure for representing number must know not only the how-to-count and what-to-count principles but also the order-irrelevance principle: that the order in which items in an array are counted is irrelevant. Further, we contend that a full appreciation of this principle depends on the recognition of some other implicit features of the counting procedure. In order to realize that objects in an array can be rearranged and still be counted, one must realize that the assignment of a given numerlog to a given item in an array is temporary. It will not do to think that the act of calling a particular object *one* is akin to the act of giving it a name. Some have suggested that the preschooler believes to the contrary (Ginsburg, 1975). They contend that when the preschooler counts a set of objects the tagging of each object amounts to the naming of each object. Such an argument implies that when asked to count a set with the object called *one* removed, the child should start the counting with *two.* If asked why not *one,* the child should respond that *the one* has been removed.

If young children really behaved this way, they would be violating the stable-order principle. On one trial they would use a list of tags starting with the word *one,* and on another trial they would use a list starting with the word *two.* Further, they would presumably have a difficult time recognizing that the cardinal numerosity of the set remains the same no matter which item receives a given tag. In short, the absence of an understanding of the order-irrelevance principle could be taken as evidence against our view that young children not only can apply the counting procedure but also can do so to achieve a numerical representation of a set. The child who can honor the order-irrelevance principle is therefore seen as one who has some knowledge of what the counting procedure does for him.

In several of our studies we have asked children to count the number of items in a given set several times. Usually we rearranged the items, sometimes by rotating an array so as to reverse the left-to-right placement of the items, sometimes pushing the items around in a haphazard fashion. At other times we left the objects in the same arrangement. By recording the particular object that was tagged with a given numerlog in each trial, we could look across count trials for any tendency to assign a given object the same numerlog trial after trial. One might expect that after labeling a certain object *one* on the first trial a child with no trace of the order-irrelevance principle would persist in calling the same object one across further trials, even though the object kept being moved around. In our analysis of counts across rearrangements, we seldom found any evidence of children attempting to keep giving the same numerlog to a particular item. The finding was true no matter what the age of the child. If a child showed any such evidence, it always appeared under one condition. When an array was rotated 180 degrees an occasional child would shift his starting point from the left (or right) to the right (or left). Thus in general we found no evidence that young children treat count words as names for particular objects. Indeed, they appeared to be oblivious to the order in which items are tagged.

We call this evidence on the use of the order-irrelevance principle *weak* evidence because of this apparent obliviousness. The data are not like those on which we rest our case about the use of the other principles. Our conclusion that the how-to-count principles govern the child's behavior was not based only on conventional evidence. Much emphasis was put on the use of idiosyncratic lists. For such errors reveal the presence of a rule-governed behavior, a behavior that is guided by a rule system like the one we characterize in the principles. Our concern about making too much of the weak evidence reflects our uneasiness about the possibility that the child's indifference to the order in which he tags items is just that—a case of indifference. We prefer to have data showing that the indifference reflects a principle or rule, which dictates that any item may receive any of the tags.

To obtain such data we ran experiments designed to clarify the extent to which the child appreciated or recognized this principle.

Our *strong* data on the order-irrelevance principle come from experiments in which we told children to modify the way they typically count. When children count items in a linear array, they invariably count left to right or vice versa. In the "doesn't matter" experiments we took advantage of this fact. After a child counted once, the experimenter pointed to an object in the second (or third or fourth) position and asked the child to count all the objects again making that object number one. Next the child was asked to make that same object be the two, the three, and so on. Thus we forced the child to reassign numerlogs systematically. We found that 4- and 5-year-olds were remarkably good at this task. Even some of our 3-year-olds did well. Not too surprisingly, skill at this task depended on skill in applying the how-to-count principles. The task demands that the child be able to count. That is, whatever strategies the children adopt to solve the problem posed by the task must be consistent with the counting procedure. Therefore it makes sense that children who succeed at it can likewise apply all three how-to-count principles. It also makes sense that some good counters have trouble with the "doesn't matter" task. For the task assesses more than the ability to use the counting procedure and a tendency to be indifferent as to which item receives which tag. It also assesses the child's ability to understand somewhat bizarre instructions and to generate appropriate solution strategies.

When we analyzed the relationship between counting ability and success on the order-irrelevance experiment we treated three goups of children as good counters: first, those who applied the how-to-count principles without error; second, those who applied the how-to-count principles in concert but made some errors (the sloppy counters); third, those who applied all three how-to-count principles but only on set sizes smaller than the set size employed in this experiment. To receive high grades on the order-irrelevance task a child could be at any of these levels. That is, being a sloppy counter did not rule out the possibility of doing well on the task.

We think that between the ages of 2½ and 4 or 5, some conceptual development occurs concerning the order-irrelevance principle. The development is from the stage in which the child gives what we have called weak evidence to the stage in which he performs well on the task that requires explicit recognition of the principle. In the first stage the child uses the principle as it were unwittingly. In the second stage the child takes explicit cognizance of the principle. Children who do well at reassigning numerlogs to a given item are quite good at explaining why they can do what we ask them to do. They invoke the principle in their verbal accounts. They have conscious access to the principle that permits them to modify their typical way of counting.

If we are right in this interpretation, then what develops is insight about the principles that govern counting. Thus the hard data are strong evidence that the child comes to know what it is that he does when he counts. The child not only honors the order-irrelevance principle but likewise understands what counting is about.

Is it not a problem for us that some children err in applying the counting procedure? We think not. There is no reason to assume that knowing how to do something guarantees that it will be done perfectly. Many adults understand how to do complex arithmetic calculations and yet find themselves making errors. The first author is painfully aware of this fact and is ever grateful for the widespread availability of the hand calculator—especially when it comes time to balance her accounts. It is important to recognize the fact that some children who understand the counting procedure will make errors when counting. Indeed, so will adults if required to count large sets. (And the sets need not be all that large; recall our research assistant who meant to count out 19 items but occasionally ended up with one more or one less than 19.) It is quite possible to fail to find a relationship between the ability to count without error and the ability to perform successfully on other quantitative tasks. We must distinguish between perfect skill at counting and the understanding of how to count.

Subitizing Revisited

In tracing the development of the child's application of the cardinal concept we noted a decreasing reliance on counting out loud. The younger the child, the greater the tendency to count the items in a set and then repeat the last tag. These younger children tend to count and repeat the last tag on every trial—even when every trial involves the same set of objects. Later, children count aloud on the first trial with an array, but on further trials with the same array they simply state the last tag of an earlier count, that is, the cardinal number. Still later, children state the cardinal number without ever counting aloud. This sequence of development is dependent on both the age of the child and set size. The older the child and the smaller the set size, the more likely are we to observe the child simply stating the cardinal number of the set. The same child will count aloud when shown larger set sizes.

For us, such findings support Beckmann's hypothesis (1924) regarding the relationship between counting and subitizing (the perceptual apprehension of the numerosity of the array). Beckmann hypothesized that a child first counts a given set size and only later in his development subitizes that same set size. We view the ability to use perceptual grouping strategies as an aid in the process of abstracting a numerical representation as a higher-level ability. There is a widespread assumption that subitizing is a low-level perceptual mechanism for representing the numerical value of small set sizes (see Chapter 6 for a review of the literature). If such a low-level process exists, we have not been able to observe its workings. Even our youngest subjects apply the counting procedure when they abstract numerical representations of set sizes as small as two or three. What is more, they count aloud more often than do 3- and 4-year-olds working with comparably small set sizes. Such findings force us to redefine subitizing. We say *redefine* because we want to argue that when perceptual processes are employed in the service of making quick numerical judgments they are not low level. To illustrate, consider the way an adult might respond to a request to identify the number of dots in a six-by-six matrix of dots. He might quickly count the number of rows as six, note that the array is symmetrical, and conclude that the number of columns must be six. Drawing upon his rote memorization of the multiplication table, he would then respond that there are 36 dots. Given that this adult is able to count six items at a very rapid rate and note the symmetry of the array while doing so, it would not be surprising to find that he responds quite rapidly—at least as compared to the length of time it would take him to count every dot in the array. The fact that the array is shaped like a square is used by the adult as a shortcut to a response. Obviously, then, perceptual processes can be and are used in numerical judgment tasks (compare Beckwith and Restle, 1966). But by our analysis they are used in conjunction with a counting procedure and their use depends on the adult knowing how to use perceptual relations in making numerical judgments.

It is hardly likely that preschoolers employ such advanced grouping devices; they probably use much simpler ones. What we want to argue is that preschoolers develop an ability to use perceptual strategies as they come to be sure of the results of the counting procedure. Further, we want to argue that the shortcut methods they develop are diverse and that humans seldom rely exclusively on a direct, perceptual pattern-recognition mechanism when abstracting number.

Why do we reject the idea that subitizing is a numerical abstraction method that operates independently of the counting procedure?

With subitizing broadly defined as anything other than counting, it will of course be difficult to prove that it is not used. We wish to argue, however, (1) that data exist that are inconsistent with the notion that subitizing involves the direct perceptual apprehension of numerosity; (2) that the data on reaction time, which have provided the principal justification for assuming a subitizing process, can be explained at least as readily by assuming the use of specialized rapid counting strategies with small numbers; (3) that it is possible to account for the way young children reason about small numbers by assuming the involvement of a counting algorithm but not by assuming that a subitizing process is the only way children represent numerosities; and (4) that until someone specifies a subitizing process that does *not* involve counting at some level and can nevertheless account for the multifarious data on young children's numerical abilities, the question of a subitizing mechanism as an alternative to counting will remain moot.

The study by Gelman and Tucker (1975), reviewed in Chapter 5, showed that preschool children's number estimates were independent of item homogeneity or heterogeneity. Their number estimates were also independent of visual angle in the range of approximately from 1.7° to 25.4°. This independence of item heterogeneity and visual angle was just as true for exposure times of approximately one second as for exposure times as long as one minute. Neither of these findings squares with what one would expect if the child's estimates of numerosities were based on direct perceptual apprehensions. Both item heterogeneity and wide visual angles should impede the perceptual organization of the field. This in turn should impede the direct perceptual apprehension of numerosity, unless the one-second exposure is sufficient to allow one to obtain a gestalt-like organization of the field. We believe that one second is indeed sufficient for a clear perception of the array; it was, however, short enough to interfere with the 3-year-olds' ability to estimate the numbers 2, 3, 4, and 5 and the 4- and 5-year-olds' ability to estimate the numbers 4 and 5. Therefore, although the exposure times may well have been sufficient for clear perceptions, they were not sufficient for fully accurate estimates of numerosity, a fact that argues that estimates of twoness, threeness, and so on are not based on the apprehension of distinctive gestalten. A model that assumes the application of a counting procedure following a clear perception of the array has no difficulty in accounting for all of Gelman and Tucker's findings on heterogeneity, visual angle, and exposure time.

When we further consider that young children apply numerical reasoning principles to their representations of small numbers, it becomes evident that counting *should* play an important role in the obtaining of numerical representations. It is unclear how a noncounting, subitizing procedure could supply appropriate inputs for the reasoning process. How, for example, would a pattern-based number-abstraction procedure provide answers to the relational questions posed by the reasoning principles? If the counting procedure is used to obtain a representation of numerosity, then it is clear why children regard almost any array of three elements as equivalent in numerosity to any other array of three elements regardless of the spatial disposition of the elements within the arrays. It is much less obvious what the basis of this equivalence judgment can be if numerosity is based on pattern. What is it that all possible configurations of three elements, both linear and nonlinear, have in common that distinguishes them from all possible configurations of two, four, five and six elements?

Our perplexity about pattern-based procedures in numerical estimation increases still further when we turn our attention to the ordering relation. If we assume that the representation of threeness derives from a unique pattern element common to all arrays of three and that the representation of twoness derives from a unique pattern element common to all arrays of two, why does the system see these two nonequivalent classes of arrays as satisfying an ordering relation? In considering this question, it must be remembered that the numerical ordering relation is orthogonal to the ordering relation of physical size: The child judges a linear array of two items to be less numerous than a linear array of three items even when the two-item array is longer than the three-item array. In general, it does not appear that children (or adults) organize most of their perceptual classes in accord with size-independent ordering relations, so why and how should the young child organize his perceptions of twoness and threeness in this way? On the other hand, if the child's representations of twoness and threeness derive from a counting procedure, then the ordering of numerosities is homomorphic to the ordering of the numerons or numerlogs used in counting.

It could be argued that our inability to specify the invariant features that underly the perception of threeness is nothing but a reflection of our generally poor understanding of perception. We cannot, for example, specify the invariance underlying the perception of "treeness"; yet children are able to sort the world into trees and nontrees. Defining subitizing as the direct apprehension of number, however, does imply that some hard-to-define property of threeness makes all arrays of three objects perceptually distinct from all arrays of four objects in much the same way as all trees are perceptually distinct from all cows. If this were true, the reaction time to threeness should be about the same as the reaction time to fourness, just as the reaction time to "treeness" is likely to be about the same as the reaction time to "cowness." In other words, the subitizing hypothesis has considerable difficulty in explaining why twoness, threeness, fourness, and fiveness require progressively more time to be recognized—especially by young children (Chi and Klahr, 1975). Likewise, this hypothesis cannot explain why as children develop their number abstraction abilities they learn to deal first with two-item arrays, then with three-item arrays, then with four-item arrays, and so on. Whence comes the orderly development? What characteristic of the direct perceptual mechanism dictates that acquisition will follow an order rule? Again, one could argue that our difficulties with the perceptual apprehension hypothesis derive from the absence of a precise model of how it works and that when such a model is provided the problems we raise will disappear. Perhaps this is true. Still, a counting model can account for the points we raise.

If we abandon the definition of subitizing as a process of direct perceptual apprehension, what are we to make of the traditional data used to support the postulation of such a mechanism? Woodworth and Schlosberg's well-known argument (1954) for defining subitizing as a direct perceptual apprehension mechanism was based on the assumption that the slope of the reaction time function for numbers less than four or five was flat. Chi and Klahr (1975), however, have recently shown that this slope is not flat. Adults require, on the average, about 46 milliseconds longer to make an accurate estimate of the number two than of the number one, about 46 milliseconds longer to estimate the number three than the number two, and about 100 milliseconds longer to estimate the number four than the number three. In kindergarten children (5 years of age) these increments are much larger. These children require approximately 120 milliseconds longer to form an accurate estimate of the number two than of the number one and a further 280 milliseconds to form an accurate estimate of the number three.

It seems to us that the shallower slope of the reaction time function with small sets can be explained at least as readily by assuming the use of certain specialized rapid counting strategies in dealing with these sets as by assuming the use of a subitizing mechanism. And the difference in the slopes for adults and children can be taken to reflect differences in the extent to which these strategies have been developed and routinized.

The reader will perhaps be convinced of the possibility of such rapid counting procedures for small numbers after trying the following phenomenological experiment. Count to 3 a few times to yourself as rapidly as possible, and rate your confidence about actually having counted up to three. Then go through the same process with counting to 20. If your experience resembles ours, your confidence in the accuracy of your counting will be high when you count to 3 but will deteriorate almost to nothing when you try counting to 20. The point of this experiment is to demonstrate that it is plausible to assume that unusually rapid counting strategies can be applied more reliably to small numbers than to larger ones. As with most phenomenological experiments, it is very hard to get subjects to agree about the point at which the slowdown begins. Furthermore, we can offer no detailed descriptions of the differences in strategies. Perhaps because we lack such descriptions, we cannot explain the apparent shift in strategies that occurs in the range of set sizes from three to five. We suspect that the strategies have something to do with the rhythmic organization of action. It may be no accident that most of our musical rhythms are based on groups of no more than three or four beats.

We think that the development of rapid counting strategies for small sets begins when the counting of small sets has become largely a subvocal process. Presumably, these strategies are developing between the ages of 3 and 5. Recall that Gelman and Tucker found interactions between age, set size, and exposure time. A reduction in the exposure time to one second affected the accuracy of estimation by 3-year-olds regardless of set size; it also affected the accuracy of 4- and 5-year-olds for set sizes of four and five but not for set sizes of two and three. Gelman and Tucker reported a complementary interaction between set size and age in the data on counting aloud. Five-year-olds tended to count aloud only when confronted with larger sets. We think these data show that one consequence of the routinization of the counting procedure is that counting becomes covert rather than overt. This change in turn could permit the development of rapid counting strategies for use with small numbers.

It should be noted that nothing in our account precludes the possibility of using perceptual grouping processes along with the counting process. Indeed, we suspect that a full account of the strategies used in making absolute judgments will be quite complicated. We presume that even if the reader is unwilling to accept our hypothesis of rapid counting and its role in the development of number-abstraction skills, he will agree that the term *subitizing* must be redefined to be no longer a low-level, "primitive" way of abstracting the numerosity of sets.

ALGORITHM DEVELOPMENT: SOME GUESSES

The routinization of the counting process may also be a prerequisite for the development of recursive counting algorithms that may be used in the solution of addition and subtraction problems. As we have pointed out earlier, younger children compute the solution to an addition problem by counting the union of two sets from beginning to end. Thus, when asked "how much is two and three," these children count out a set of two (for example, two fingers), then count out a set of three and finally count the union of the two sets. With development, the answer is arrived at more quickly. The child starts his computation with the numerical representation of one of the sets, either three or two, and counts up from there with a number of count steps equal to the numerosity of the other set (Groen, 1967). This presumably involves a recursive use of the counting procedure even when, as is not uncommon, the child uses his fingers as props. Thus the child may start by saying "three" and count "four, five" while raising two fingers. The recursive use of the counting procedure is implicit in the child's knowing to stop at five, that is, after two steps in the counting process. The realization that the developmentally later computational strategy involves the recursive use of the counting procedure helps explain the developmental sequence. One would not expect to see the recursive use of a procedure before the procedure itself was fairly thoroughly routinized.

In subtraction problems, as in addition problems, the answer may be computed either recursively or nonrecursively. In subtracting three from five, the child may compute the answer by counting and removing three items, then counting the remaining items. Or the child may start with the smaller number and count up to the larger number and report as his answer the number of counting steps. Note that in the recursive solution to subtraction problems, the number of counting steps is the unknown whose numerical representation is being computed. Both the starting point and the stopping point are known. In the recursive solution to addition problems, the starting point and the number of counting steps are known, and the stopping point is the unknown whose numerical representation must be computed.

Many problems that require subtraction do not lend themselves to solution by the simpler nonrecursive method. This is particularly true of problems in which the child is required to compute a numerical representation of the difference between two sets, that is, to solve for the difference. In this situation, the problem *as stated* does not involve removing a proper subset and counting the remainder. Thus the child must tackle the problem recursively or translate it into a related problem that does involve the removal of a proper subset. In either case, a "solve for the difference" problem should be more difficult than a "take away" problem. Schools introduce "take away" problems before difference problems; presumably this order reflects the fact that educators consider "take away" problems easier than difference problems. In our own research, the children were confronted with a difference problem whenever they were asked how many things had been removed or added. The same problem was presented in a practical guise whenever they were asked to fix the game, that is, to restore the winner set to its original numerosity. The children were accurate when solving for a difference of one. Solving for a difference of two, however, seemed to tax the computational abilities of most of the children. They knew that the required answer was more than one (as evidenced by their use of plurals), but they seemed unsure of its exact numerical representation. For the most part, children fixed the game by trial and error. Thus for example, many children who were required to make a three-item plate into the original five-item plate began by placing a number of objects on the plate. They then proceeded to count, add or remove one, count, add or remove one, and so on until they had a set of five objects. At this point, they stopped and said they had the winner plate.

In summary, the computation of the appropriate numerical representations in many addition, subtraction, and difference problems involves an algorithm that uses the counting procedure recursively. It seems reasonable to assume that this recursive use is not practical until the procedure itself is routinized, that is, fast, efficient, reliable, and requiring little attention. The development of these recursive algorithms expands the child's domain of computational competence.

Verbal versus Nonverbal Activation of Numerical Processes

Those who attempt to question children of preschool age directly about their numerical knowledge generally find that the children are not very responsive, nor do their answers seem to indicate much numerical knowledge (see for example, Rothenberg, 1969; Siegel,

1976). Yet the magic experiments indicate a fairly sophisticated set of numerical reasoning principles and some modest computational skills on the part of children as young as 2½ years. The magic experiments were specifically designed to confront the child with an indirect (not verbally stated) numerical problem in the midst of a game that captured the child's interest. The considerable discrepancy between our data and data from experiments that pose numerical questions verbally highlights another important line of development—an increase in the facility with which information given or requested verbally can access the arithmetic domain. This line of development seems to involve two quite distinct problems.

The first problem concerns the mapping of the linguistic domain into the cognitive domain. Many words that refer to aspects of a child's numerical reasoning principles are very ambiguous because their use is not confined to the numerical domain. This is particularly true of the relational terms *more, less, same as,* that figure so prominently in numerical questions directed to children. One could hardly find terms with a wider domain of possible reference. The child must grasp the multidimensionality of these terms and learn to derive from context the intended dimension in a given situation. Children appear to have more difficulty with these terms than with the unambiguous phrase *how many.* We have little difficulty running 3-year-olds in absolute estimation tasks using *how many.* In contrast, Rothenberg (1969), Donaldson, and Wales (1970), and others report that 3-year-olds are unable to respond accurately to questions asking for judgments of *more* and *less.* The data from the magic experiment clearly show that this failure does not derive from the absence of an appropriate ordering relation in the child's numerical reasoning principles (see also Siegel, 1974).

The posing of computational questions verbally raises the second problem of this line of development. When the child is given the number-word problem "how much is three take away two," either he must compute the answer in his head and therefore through the use of a recursive algorithm or he must set up or find his own props. As we have already discussed, the setting up of props—generally fingers —is a developmentally earlier stage in the solution of number-word problems.

Reasoning Principles

Toward Algebraic Reasoning about Number

In the Piagetian treatment of the development of numerical reasoning, it is assumed that the child's ability to treat two sets as numerically equivalent rests on his belief in the existence of a one-to-one correspondence between two sets. For Piaget, then, one-to-one correspondence is the psychologically primitive basis for a judgment of numerical equality. One attraction of this view is that one-to-one correspondence is almost invariably taken as the definition of numerical equality in attempts to construct the number system from set theory or the predicate calculus (see Kline, 1972). Thus, in the Piagetian view there is a parallel between what is taken as primitive in formal developments of arithmetic and what is psychologically primitive. We believe that this parallel does not hold. Although we agree that at some point in an individual's development he recognizes that sets that can be placed in one-to-one correspondence are equal, we do not agree that this point marks the emergence of the initial appreciation of numerical equivalence. Instead, we argue that the ability to make judgments of numerical equivalence based on one-to-one correspondence is a later stage in the use of reasoning principles.

We find that the preschool child's judgment of whether two sets are numerically equal ordinarily rests on whether they yield the same cardinal numeron when counted. In other words, to the young child, two sets are numerically equal if the application of the counting procedure yields the same numerical representation for both sets. The evidence for this conclusion comes from two parts of the magic experiment. First, children judge an array of n items to still contain n items after the length, color, and identity of items within the array have been altered. Second, when children are asked to undo the changes in the array, they often construct two sets that have the same numerosity, especially if this means that they will end up with two winners. It is important that in neither of these situations do the children confront sets that are placed one above the other to make the one-to-one correspondence obvious. In the first situation they compare the number of items in a present array with a memory of the number of items in an earlier array. In the second, they are comparing two arrays that are placed not in one-to-one correspondence but rather side by side.

We draw attention to these facts to highlight a feature of Piaget's conservation task, namely that the two arrays that are to be compared *are* placed one above the other. Initially the sets are displayed so that the one-to-one correspondence is obvious; then one array is lengthened or shortened, destroying the perceptual correspondence. Young children first say that the sets contain the same number of items but later say that they do not. Despite the fact that they do not see the experimenter add (or subtract) items to (from) either array, they maintain that one array has more (fewer) items than the other. This result is a puzzle. It cannot be as Piaget says (1952), that the child lacks a concept of number or numerical reasoning principles. For we have demonstrated to the contrary. From our viewpoint, the problem is that the child fails to *use* his reasoning principles. What we need is an explanation of why he does not apply his principles.

We take the position that initially the young child reasons about number only when he can obtain a specific consistent numerical representation of a collection. Can this assumption lead us to an account of the young child's failure on the conservation task? We have already considered one possibility (see Chapter 8). Recall that young children have a strong tendency to make partitioning and coordination errors when counting sets of seven or more items. The conservation task typically involves set sizes of this magnitude. Given that the young child defines equivalence in terms of a counting procedure, the only solution strategy open to him is one based on counting. But if he counts, he will probably err on at least one of the two critical counts. This explanation of failure on the conservation task requires that the child count and err in doing so. This is certainly a possibility, but it cannot be the whole account, simply because young children in the conservation experiment seldom count spontaneously. Further, this explanation requires that counting be observed on the part of children who do conserve. Again, counting does not seem to be prevalent behavior (Miller and West, 1976). Whatever relationship exists between counting and conserving is not a strong one. Many children fail to conserve even when run on the conservation test with small arrays that they can count easily and reliably. And it is possible for children to make counting errors and still conserve (Saxe, 1977). For all these reasons, it makes little sense to explain the ability to conserve as a simple extension of counting skills.

Our explanation of why the young child fails to use his reasoning principles centers on those children who use one-to-one correspondence to form an initial judgment of equivalence. We set aside the children who begin by focusing on the common length of the arrays. The child who does use one-to-one correspondence has, from the adult point of view, made a numerical judgment. Therefore the question of why he does not conserve becomes the question of why he gives up his numerical judgment. The child who starts with the idea that the length of the arrays is what matters is not even treating the task as a problem about number.

We believe that the child who begins by responding on the basis of one-to-one correspondence fails to apply his reasoning principles because the perceptually obvious correspondence creates two problems. First, it induces a judgment of equivalence in a form that cannot be dealt with by the child's reasoning principles. In their initial form these principles do not apply to equivalence relations between unspecified numerosities; they apply only to equivalence based on the identity of the cardinal numerons. Second, the salience of the pairings between corresponding items in the two rows impedes the application of reasoning that avoids the first problem, namely reasoning based on the cardinal values obtained by counting each row. The between-row pairings distract the child's attention from the within-row numerosities that are the focus of the experimenter's interest. The pairing impedes reasoning based on the cardinal values obtained by counting each row by distracting attention from the actual numerosities of the rows or the fact that there are two rows, each of which represents a numerical value.

We suggest that this second problem may continue to impede correct conservation at an intermediate stage in the development of the child's reasoning. In this intermediate stage, the child's reasoning about number-irrelevant transformations applies to a single unspecified numerosity but still does not apply to an equivalence relation between two unspecified numerosities. In Elkind's terms (1967), the child conserves numerical identity but not numerical equivalence. In this stage the child can arrive at a correct conservation response by a two-step chain of reasoning (described below) without first obtaining representations of the specific numerosities involved. The salience of the pairings, however, may keep the child from pursuing the line of reasoning that leads to the correct response. As in the earlier stage, the pairings draw attention away from the fact that each row constitutes a numerosity.

In the final stage of development, the stage of mature or full-fledged conservation, judgments of numerical equivalence based on one-to-one correspondence no longer pose a problem because the child's reasoning principles apply directly to equivalence relations between unspecified numerosities. Also, at this more advanced age the child is unlikely to overlook the experimenter's evident interest in the numerosity of the rows (Gelman, 1969). Thus in this stage, which we call the stage of algebraic reasoning, the pairings no longer offer any sort of impediment to reasoning.

This argument assumes that arithmetic reasoning develops from a numerical stage to an algebraic stage. Numerical reasoning deals with representations of specific numerosities. Algebraic reasoning deals with relations between unspecified numerosities. To illustrate this distinction, consider what the prealgebraic child can and cannot do. He can count one set and get the number four; he can count another set and also get the number four; he can judge the two numerosities to be equal because they are both represented by the number four. He believes that displacing one does not destroy its equivalence to the other because the numerosity of the displaced array will still be represented by the number four. In other words, he believes that the representation of a numerosity is not affected by displacement. He recognizes the countinuing identity of four to four. Now consider his dilemma when he judges two sets of four items each, placed in one-to-one correspondence, to be numerically equivalent on the basis of the one-to-one correspondence rather than on the basis of the identity of the cardinal numerons. When asked to reason about the effect of displacing one of these numerosities, he has no appropriate input for his reasoning principles: no specific representation of the numerosity. He has only a representation of the algebraic form $x = y$, where x and y represent unspecified numerosities. But he does not have an algebraic reasoning principle of the following form: The equivalence between two entities, x and y, representing unspecified numerosities, is unaffected by displacing either the numerosity represented by x or the numerosity represented by y. We call a principle of this type algebraic because it applies to entities (x and y) that represent different numerosities at different times, whereas the entities (numerons) to which the young child's principles apply always represent the same numerosities. The numerosity represented by x may be numerically equivalent to that represented by y at one time and not at another time. But a numerosity represented by four is always equivalent to any other numerosity represented by four; it is never equivalent to a numerosity represented by three.

When number comes to be viewed algebraically, the focus of attention is no longer on number as such but rather on numerical relations. Algebraic reasoning principles manipulate systems of numerical relations, not a system of specific numbers. The x's and y's that stand for the now-vanished specific numerosities are dummy variables. The relation itself is the true variable: The dummy variables x and y are there only because the relation of equality implies that there exist two entities between which the relation holds. Similarly, when we want to focus on the operation of addition, rather than on the sum of two numerosities, we need dummy variables that stand in for numbers but do not represent any specific numerical value.

In sum, we are arguing that young children are unable to reason about unspecified numerosities. Even if a young child notices the one-to-one correspondence between the items in two arrays, he still will fail the conservation test. His reasoning principles do not tell him how transformations affect relations; they tell him how transformations affect numerosity. We further hypothesize that the child who does conserve has developed the ability to reason about numerical relations in the absence of representations of specific numerosities. This is another way of stating the hypothesis that what develops is the ability to work with algebraic inputs rather than just numerical ones.

As indicated, our account of the young child's failure on the conservation task focuses on the child who attends to the one-to-one correspondence and the numerical equivalence it implies. We need evidence that young children can and do take note of the item-by-item correspondence between two sets and still fail to conserve. Such evidence is provided by Piaget (1952). Piaget had children construct an equivalence on their own. The children were given one row of objects and asked to place another one above it so that both rows contained the same number of items. Piaget reports that some children succeeded at this task by placing an item directly above (or below) each item in the standard display and still denied that the two arrays were equivalent when one was displaced.

We have explained why a preschooler at the first stage of development who understands the experimenter's words and who attends to the property of the display that the experimenter wants him to attend to—the one-to-one correspondence—fails the conservation test. We have not explained why such preschoolers often fail the test even when conditions permit them to solve using abilities they clearly possess. Many children at this stage do not conserve even when tested with arrays small enough for them to count. Making the arrays small enough to be easily counted increases the percentage of young children who conserve, but still many do not. This failure to employ alternative lines of reasoning we attribute to the second problem created by the salience of the one-to-one pairings in the first phase of the test. These between-row pairings draw attention away from the rows themselves and therefore away from properties of the rows, particularly their numerosity.

One piece of evidence that the pairings divert attention from the rows is Miller and West's finding (1976) that the more one emphasizes the one-to-one correspondence the less counting one observes. Counting is a procedure applied to the entire row in order to obtain a representation of a property of the row. It is not surprising that emphasizing the pairings, which constitutes a different way of structuring the array, impedes processes that demand a focus on the row structure.

The most persuasive evidence that the degree of attention to the row and its properties is an important variable in conservation comes from experiments now being conducted by Markman. We are grateful to her for sharing her data and ideas with us. Markman modifies the standard conservation question to induce children to think of the rows as collections and not as classes. Thinking of the rows as collections considerably increases 4-year-olds' tendency to conserve.

The work is an extension of her studies on the difference between collection and class concepts. Markman and Siebert (1976) call attention to the fact that many important concepts do not follow the logic of classes. Classes are defined by their intension and extension. The intension of a class is the criterion or principle by which one decides whether something does or does not belong. The extension of a class is all those entities that meet the intensional criteria. The intension of the class of preschoolers is "a human being less than 6 years old." The extension of this class is all human beings who at this moment actually are less than 6 years old, that is, all things that are instances of human beings less than 6 years old. In the logic of classes one can decide whether or not something is a member of a class simply by considering that something; one does not have to consider the relation between that something and other somethings.

A collection, on the other hand, is defined by the existence of a particular pattern of relations between its members. One cannot decide whether something is a member of a collection simply by considering that something. Instead, one must consider the relations of that something to other potential members of the same collection. Consider, for instance, the collection that constitutes a family. A child is a member of such a collection only by virtue of its relations to other members. Whether a child is a member of any family at all depends upon whether there exist other human beings who are relatives of that child—mother, father, brother, sister. If it is not known whether other human beings exist who have the required relation to a child, it is not possible to decide whether the child is a member of a collection we call a family. Other instances of collections are an army (as distinct from the class, soldier), a faculty (as distinct from the class, professors), and a pile of bricks (as distinct from the class, bricks). Notice that the logic of classes permits one to focus on individual members whereas the logic of collections requires attention to the collection itself and the emergent properties of the collection. Number is just such an emergent property. Numerosity is a property of a collection or class as a whole, not a property of its individual members.

Markman hypothesized that inducing children to think of a row as a collection rather than a class would direct their attention to emergent properties such as numerosity. She showed children toy soldiers arranged in two rows just as they would be at the start of the conservation experiment. One group of children was asked a collection question: "Does your army have as many as my army?" The other group was asked the standard question, which happens to be a class question: "Do you have as many soldiers as I do?" Markman reasoned that the collection questions would elicit more conservation judgments than the class question, since class questions focus attention on individual items whereas collection questions focus attention on all the items that together make up a collection. Number is a collective attribute; it applies to the collection or set of items as a whole and not to the individual items.

Children were tested with either collection or class questions, and each child was given four conservation trials. Those tested on class questions answered about one and one-half of the four items correctly. Those tested on collection questions did twice as well. Children in the collection condition not only gave more posttransformation judgments of "same" but were able to give explanations as well.

From our point of view, Markman's collection questions served to emphasize the integrity of each of the two sets of objects. The continued use of such questions made it possible for the children to focus on the fact that the number of items in each collection remained unchanged.

The argument we are developing has two components. The second component concerns the effect of salient between-row pairings on the child's tendency to employ lines of reasoning that do not require dealing with an equivalence relation between unspecified numerosities. The first component concerns the stages that a child's reasoning passes through en route to the final stage, the algebraic stage, in which he can finally deal directly with equivalence between unspecified numerosities. The kinds of justifications Markman's 4- and 5-year-old subjects gave for their conservation judgments reflect the two earlier stages. Some children justified their judgments by mentioning the actual numerosities and calling attention to the fact that they were the same. In other words, they simply pointed out that both rows still had, for example, seven items. This kind of justification reflects the earliest stage, the stage of purely numerical reasoning, the stage that requires specific numerical values. Other children simply stated that the transformation was irrelevant to number (for example, "you just moved them"). We believe that such justifications reflect a semialgebraic stage.

Children in this intermediate, semialgebraic stage can reason about the effects of transformations on a single unspecified numerosity. They cannot reason about the effects of transformations on the relation between two unspecified numerosities. Children in this stage may or may not base their initial equivalence judgment on one-to-one correspondence. Whether or not they do so, they have open a line of reasoning that will lead to a correct response on the conservation test. They can reason that the transformation did not alter the numerosity of the transformed set. The numerosity of the nontransformed set, of course, remains constant. So the two unaltered numerosities must still be equivalent. It is important to recognize the subtle but significant way in which this line of reasoning differs from the stages before and after it. It differs from the stage before it in that it does not rest on the preservation of the identity of the cardinal numeron representing numerosity. The reasoning works whether or not one has a representation of numerosity; thus we call it semialgebraic. The reasoning is not fully algebraic, as in the next stage, because it does not deal directly with the irrelevance of the transformation to a relation between unspecified numerosities. The children do not justify their judgment by mentioning either the original equivalence or the continuing existence of a one-to-one correspondence. They mention only the irrelevance of the transformation to the numerosity of the transformed array.

We summarize our argument using the notation introduced in Chapters 10 and 11. Initially, the young child's reasoning principles can only be applied to specific numerical values. In other words, there exists a class of operations, I, for which $C[I(S_n)] = C(S_n)$. This notation represents the principle that operations belonging to the class I of identity operations transform a set of numerosity n into a set of numerosity n. The young preschooler's version of this principle requires that $C(S_n)$ have a specified value. A set whose numerosity is not actually represented by a cardinal numeron is a set to which this principle cannot be applied as far as the very young child is concerned.

The first step toward algebraic reasoning occurs when the child applies this principle in the absence of a specific value for $C(S_n)$. Now the child draws the inference represented by the notation $\tilde{C}[I(S_n)] =$

$\tilde{C}(S_n)$, where $\tilde{C}(\ )$ represents the cardinal numeron that *would* be obtained *if* one counted the set. In this stage the child can recognize the numerical invariance through an I transformation of an array that he either cannot count or has not counted. To do this, however, he must attend to the fact that the array has a numerical value, whether or not he represents that value.

The second step toward algebraic reasoning occurs when the child recognizes that transformations from class I do not affect numerical relations. When this stage is reached, the reasoning principle applies to relations between numerosities rather than to individual numerosities. The reasoning principle might now be formalized as follows: If $\tilde{C}(S_x) = \tilde{C}(S_y)$, then $\tilde{C}[I(S_x)] = \tilde{C}(S_y)$. Because this principle does not require that the counting procedure be performed, and because this principle focuses on an arithmetic relation rather than on numerosity, we say the child who reasons thus has reached the algebraic stage.

We call attention to the conceivable existence of another stage of development, for which one might write, if $N(X) = M(Y)$, then $N[I(X)] = M(Y)$, where $N(\ )$ and $M(\ )$ represent the numerical values assigned to the *things* specified in the parentheses and I symbolizes an operation that may be performed on a thing without altering the numerical value assigned to that thing. For example, I would symbolize all possible operations that can be performed on the energy of a closed physical system, since nothing can alter the numerical value that is assigned to that energy by the operation of measuring that energy. In this stage, arithmetic reasoning is no longer limited to dealing with representations of numerosity. In now deals with that ethereal abstraction called number. We doubt that this final stage is ordinarily achieved in the absence of extensive and explicit instruction.

Concerning the implications of success on the conservation task, a distinct kinship exists between our views and those of Piaget (1952). Piaget's notion of an operational concept of number and our notion of an algebraic treatment of number are closely related. Both imply a freeing of numerical reasoning from dependence on assessments of the actual numerosity of the given. Our views and Piaget's part company with respect to the implications of failure to conserve. Piaget regards the failure to conserve as a sign that the child lacks a concept of number, that is, a coherent set of principles for reasoning about number. We, on the other hand, have argued throughout this book that the preschooler has a coherent set of principles for reasoning about numerosity. The knowledge that children reason cogently about specific numerosities long before they pass the conservation test forestalls the conclusion that "preconservers" lack a concept of number. We are led instead to the conclusion that what such children lack is the ability to reason about numerical *relations,* that is, the ability to reason algebraically.

Other Operations

Our account of the preschoolers' numerical reasoning principles deals with the operations of addition, subtraction, and solving for a difference but not with the operations of multiplication and division. Since we have yet to research the question of whether preschoolers use multiplication or division operations, we must remain open to the possibility that they do. We venture to guess that if any preschoolers do employ such principles, it is the older ones. Many reasons exist for assuming that the understanding of multiplication and division comes after the understanding of addition, subtraction, and solvability.

We conjecture that the multiplication operation is slowly introduced into the numerical reasoning scheme through a long and variable developmental course that reflects an intricate interplay between endogenous and exogenous developmental forces. The endogenous seeds of this operation are to be found, in part, in the demands the counting procedure makes on memory. The counting procedure requires the storage in memory of an indefinitely long, stably recallable list of distinct numerlogs. This is precisely the kind of storage for which human memory, as opposed to, say, computer memory, is distinctly ill suited. The universal response to this problem in systems of enumeration has been the adoption of a set of tag-generating rules that represent large numerosities as products and sums of smaller numerosities. Forty represents, and obviously derives from, the expression "four tens," that is, the product of four and ten. Forty-three is "four tens plus three." This method of generating the tags for higher numerosities makes implicit use of the operation of multiplication. Thus, one could argue that the seeds of the multiplication operation are sown by an endogenous conflict—the conflict between the requirements of the counting system and the limitations of memory.

Another feature of counting also seems to lead toward the recognition of a multiplication operation. In many cultures, individuals are confronted with the task of repeatedly counting large sets, such as the number of cattle in a field. But the larger the set size one has to count, the greater the likelihood of making an error or of losing one's place. What to do? Break up the set into smaller set sizes that can be counted accurately and then count the number of sets containing x items each. If one is counting pennies one by one rather than, say, by groups of five, it is intensely annoying to have to start all over again when one has lost one's place at 79.

We think that the extent to which the seeds of multiplication germinate and grow depends on exogenous pressures, particularly on the extent of numerical work that the cultural environment requires of the individual. Children who live in environments that contain schools will inevitably be confronted with the operation of multiplication. They will be tutored in the facts that we have outlined above. Thus they will be told in more or less explicit terms that twenty-three is two tens plus three, but one hundred is ten tens, and so on. And in many cases they will be introduced to multiplication with lessons that teach them that a set of eight objects contains two subsets of four objects and four subsets of two objects. Many curricilum materials are based on what we have labeled the endogenous pressures that lead the individual in search of multiplication.

It is not just those who go to school who will discover the operation of multiplication. Where the members of a culture are required to repeatedly count large sets, in the case of cattle herding, they invariably make up something like multiplication. They typically make their counts with the assistance of props, such as hatch marks or pebbles. And these counts are usually broken into groups of shorter counts. An excellent example is given by Zaslavsky (1973) in her discussion of cowry shell counters in Africa. Cowry shells were a highly inflated but common currency in nineteenth-century African trade. The small value of the currency required the counting of thousands of cowry shells for even rather small transactions. The cowry counters performed their counts by pulling out groups of six cowry shells at a time. (It seems likely that groups of six were counted because they were within the range that is possible to subitize, or count rapidly. European traders commented on the rapidity and accuracy with which the professional counters could pluck out six shells.)

Counting precounted groups results in two numbers, one representing the within-group count, the other representing the between-group count. The determination of the single number that represents the product of these two numbers is multiplication. The algorithms for carrying out this determination are complex and various (see Kline, 1972). Historically, the extent of their development in a given culture has been influenced by the importance of trade in that culture (Zaslavsky, 1973). In modern times, it is a good bet that the extent of their development in an individual within a culture depends on the availability of schooling.

In summary, we consider the developmental course that leads to the understanding of multiplication to be a particularly salient example of how the numerical reasoning principles will accommodate to environmental pressures. We think that the expansion in conceptual ability that occurs when the reasoning principles take in the operation of multiplication is largely the result of practical difficulties encountered in the use of a simpler scheme. Given a cultural context that encourages individuals to make large and accurate counts, the situation is ripe for individuals in that culture to discover or learn about the operation of multiplication.

We hesitate to discuss the development of the understanding of division. To be sure, it is related, even dependent upon, an understanding of multiplication. But we have confessed to knowing little about the psychological makeup of an understanding of multiplication. Therefore, it seems prudent to resist the temptation to discuss division and to await the data on multiplication. We do note however, that since time immemorial the operation of division has posed considerable difficulties even for talented mathematicians. Division can be thought of as the inverse operation of multiplication. Since the inverse of almost any operation is more difficult to deal with than its source operation (Kline, 1972), it would not be surprising to find that the understanding of division follows the understanding of multiplication. Such an assumption seems to be made by those who design curricula for the teaching of mathematics. Lessons about division are introduced after lessons about multiplication. In drawing attention to such facts, we do little but restate what is obvious in the historical record and in mathematics texts. Nothing is conveyed about the psychology that might govern the understanding of division. This remains a subject for further research.

Algebraic Pressures

In accounting for success on the conservation task, we made much of a distinction between algebraic reasoning and specific numerical reasoning. The idea was that the child who succeeds at the standard conservation task has developed the ability to reason about numerical relations without reference to specific numerosities. The child no longer needs to count in order to reason about number. We think that passing the conservation test marks the onset of an ability to reason numerically with representations of unspecified or unfixed numerosities. Some evidence in support of this view comes from pilot data collected by Evans and Gelman on the development of the child's understanding of infinity. Children who give clear evidence of knowing that there is no such thing as the biggest number are able to pass the number-conservation task. The nature of their explanations is most revealing. In essence they argue that for any given N, there is always an $N + 1$. In other words, they articulate a principle of iterative addition. Being able to conserve number does not guarantee the development of an understanding of the concept of infinity, however. Some children conserve and yet insist that there is some largest number, some number to which the $N + 1$ rule does not apply.

The development of the concept of infinity is being investigated now by Gelman and Evans. One thing that seems certain already is that it need not involve formal schooling. The Gelman and Evans sample includes children in the first and second grades of a typical middle-class school. No mention of infinity can be found in their math texts, nor is there any evidence of teachers attempting to cover the topic. Yet during their participation in the experiment some first and second graders seem to *discover* that there is no such thing as the largest number. Participation in the experiment involves, among other things, answering questions about the effect of adding one to large numbers. Such questions take the following form: "What happens when you add one to ___?" "What happens when you add one to that?" "What happens when you add one to *that*?" "Is there a largest number?" "Why?" This format leads some children to discover the concept of infinity. They obviously need some environmental input, but it does not have to be in the form of instruction.

We suspect that there are other cases where the ability to reason about unspecified or algebraic representations of number can develop without the benefit of formal schooling. Indeed, the rule of equivalence based on one-to-one correspondence has been used in many an unschooled culture (see Chapter 7). Likewise, it is clear that children of all cultures come to conserve whether or not they are schooled (Cole and Scribner, 1974). We do not want to give the impression, however, that the child who conserves will develop into an algebraist on his own. Conservation may mark the beginning of this development, and some development may take place without formal schooling. But there clearly have to be limits to the latter possibility.

We think that formal instruction is necessary for the development of a true understanding of zero as a number. In Western culture zero was not treated as a representation of a numerosity until the Renaissance. Mayan culture seems to have been ahead of Western culture in this regard, but the historical record nonetheless makes it clear that the human mind is loathe to include zero with the other representations of numerosity. Our account of the development of numerical abilities provides a ready explanation of this reluctance. Developmentally, numerical reasoning principles at first apply only to representations of numerosity derived from counting sets. The counting of sets cannot yield any representation of sets of numerosity zero, that is, any representation of the empty set. Recall that counting requires the withdrawal of tags in lockstep with the transfer of items from the to-be-counted category to the already-counted category. The empty set has no item to be transferred; hence it is impossible to both withdraw a tag to represent the numerosity of the empty set and remain faithful to the rules governing the counting procedure. The historical record suggests that this psychological obstacle to the representation of the empty set was overcome eventually by pressures arising from the algebraic use of the numerical reasoning principles. Algebraically, the subtraction of some numerosity y from some numerosity x yields some numerosity z. Whenever we specify actual values for x and y we can also specify an actual value for z. However, unless we let something represent the numerosity of the empty set, we cannot specify an actual value for z in the case where $x = y$. Thus it seems natural to regard the empty set as having a numerosity if we view that numerosity as being the consequence of subtraction but not if we view numerosity as a property of countable sets. This may suggest how one could most readily teach the child the concept of zero.

The case where $y > x$ presents even greater difficulties when it comes to assigning a numerical value to z. Here, the algebraic representation of abstractions that may be considered to have numerosity becomes an important source of endogenous pressure on the system of numerical reasoning. The importance of abstractions that have numerosity only because the mind chooses to treat them as though they had numerosity stems from the fact that it is physically impossible to take away seven items from a set that has only five items to begin with. There is no physical set with numerosity negative two. However, once the mind has created an abstraction such as "my net worth," an abstraction the mind regards as having a numerosity, then the mind can respond to the algebraic pressure that arises from evaluating the expression $x - y$ in cases where the numerosity assigned to y is greater than the numerosity assigned to x. It appears that negative numbers were introduced into Western mathematical culture in response to the mathematician's need for algebraic generality (Kline, 1972) and to the emergence of concepts such as net worth in the conduct of trade, concepts that created numerically valued variables that one could think of as having a negative numerosity without doing violence to one's intuitions of what was physically possible (Zaslavsky, 1973). Again this suggests how school children might be most readily induced to incorporate negative numerosities into their numerical schemes.

It would be grand to be able to specify how children develop an abstract attitude toward number, how they progress through the acquisition of algebraic concepts, and what types of inputs support these developments. We do not pretend to know the answers to these questions. Nor are they the questions we set out to answer. We sought to determine whether young children can and do reason about number. We do know the answer to this question. We also sought to determine whether the evidence about the preschooler's arithmetic abilities would shed light on the nature of how the child's concepts of number and ability to reason numerically develop. We do not pretend to have provided definitive answers to these questions. We have provided hypotheses that can be tested, and we believe that at least some of them will continue to be supported by the experimental data.

Much of what we have been able to say about development in this chapter derived from our knowing what young children can do as well as what they cannot. The discoveries we have made about development would not have been possible if we had followed the trend of considering preschoolers merely as beings who lack the capacities of their older siblings. Our hypothesis of more capacity then meets the eye has served us well. We expect that researchers who keep their eyes open will find still more unexpected ability in young children.

Conclusions

The work reported here shows the importance of investigating both the child's ability to obtain representations of numerosity and the child's ability to reason arithmetically.

The young child obtains representations of numerosity by counting.

The child's ability to count is governed by a set of counting principles. These principles constitute a scheme in that they both guide and motivate the development of proficiency at counting.

The counting principles are (*a*) the one-one principle, (*b*) the stable-order principle, (*c*) the cardinal principle, (*d*) the abstraction principle, and (*e*) the order-irrelevance principle.

The performances required to use the principles consist of several component processes. This is particularly true of the one-one and order-irrelevance principles. We trace the development of proficiency at each component and the coalescence of the components.

The application of the cardinal principle presupposes the successful application of the one-one and stable-order principles. This fact explains why the cardinal principle is seen only in children who show some proficiency with the other two principles.

Even very young children apply the abstraction principle, that is, count sets of diverse materials. This observation calls into question the notion that the recognition of numerosity depends upon the development of hierarchical categorization abilities.

Children's ability to apply the how-to-count principles is a function of set size. Many 2- and 3-year-olds do not count reliably beyond three or four. They will attempt to do so, however, and they clearly do not regard numerosities beyond four or five as undifferentiated "beaucoups."

Children can count without using the conventional count words or the conventional count sequence.

The data indicate that the term *subitizing* must be redefined. It should not be thought of as a low-level, developmentally primitive, perceptual mechanism that simply apprehends numerosity. Instead, it should be thought of as way to group elements together so as to enhance counting.

The reasoning principles available to the young child include a set of recognized numerical relations (equivalence and order), a set of operations (addition, subtraction, and identity), and a principle of solving for a difference.

Initially children cannot reason about numbers without reference to representations of specific numerosities. These representations are obtained by counting. The judgment of equivalence or order, the application of the operations of addition, subtraction, and identity, and the process of solving all depend on counting.

Because judgments of equivalence and order depend upon specific representations of numerosity, it is not possible to determine whether the young child recognizes the transitivity of these relations. For similar reasons, it is difficult to determine whether the child recognizes the associativity and commutativity of addition. However, the child's beliefs about the nature of addition and about the irrelevance of the order of item enumeration, when taken together, are tantamount to a belief in the associativity and commutativity of addition.

The young child's system of numerical reasoning does not constitute a *group*. It lacks an identity element (no zero), closure with respect to subtraction (no negative numbers), and closure with respect to addition (no recognition of the unending succession of integers).

The preschooler does have a concept of number—a concept that contains many of the seeds from which modern arithmetic has grown.

The seeds of the child's numerical abilities grow in different ways as a consequence of diverse developmental processes:

The development of counting skill and the routinization of counting constitute one line of development. The motivating and guiding functions of the underlying counting scheme are manifest in the process by which skill is achieved. The children make up their own lists, self-correct, and spontaneously practice. These behaviors are similar to those observed in the early stages of language acquisition.

The acquisition of the sequence of count words also involves discerning the rules for generating counting tags beyond the base series.

The child's reasoning moves from a dependence on specific representations to an algebraic stage in which representations of numerosity are no longer required. This development may be seen as an instance of the process of accessing. In the initial stage, information about numerical relations in themselves does not have access to the principles of numerical reasoning. In the algebraic stage, principles of numerical reasoning have accommodated so as to apply to information about relations themselves.

The eventual adoption of zero, negative numbers, and irrational numbers—presumably under tutelage—illustrates the process of accommodation in cognitive structure following the assimilation of new instances.

The realization of the infinite progression of numbers, which seems to require only minimal tutelage at the right time in development, is an instance of induction.

The study of the numerical abilities of the preschooler yields a rich developmental account. An understanding of cognitive development must rest at least as much upon the experimental determination of what preschoolers can do as upon the experimental determination of what they cannot do. It is not enough to characterize the preschooler only in terms of his deficiencies.

References

Adams, M. J., and B. E. Shepp. 1975. Selective attention and the breadth of learning. *Journal of Experimental Psychology* 20:168–180.

Allport, D. A. 1975. The state of cognitive psychology: a critical note of Chase, W. G., ed., *Visual information processing. Quarterly Journal of Experimental Psychology* 27:141–152.

Baron, J. 1973. Semantic components and conceptual development. *Cognition* 2:189–207.

Baron, J. Forthcoming. Intelligence and general strategies. In *Strategies in information processing,* ed. G. Underwood. New York: Academic Press.

Bearison, D. 1969. Role of measurement operations in the acquisition of conservation. *Developmental Psychology* 1:653–60.

Beckmann, H. 1924. Die Entwicklung der Zahlleistung bei 2–6 jährigen Kindern. *Zietschrift für Angewandte Psychologie* 22:1–72.

Beckwith, M., and F. Restle. 1966. The process of enumeration. *Psychological Review* 73:437–444.

Beilin, H. 1968. Cognitive capacities of young children: a replication. *Science* 162:920–921.

Beilin, H. 1971. The training and acquisition of logical operations. In *Piagetian cognitive-development research and mathematical education,* ed. M. F. Rosskopf, L. P. Steffe, and S. Tabach, pp. 81–124. Washington, D.C.: National Council of Teachers of Mathematics.

Beilin, H. 1975. *Studies in the cognitive basis of language development.* New York: Academic Press.

Bereiter, C. 1968. *Arithmetic and mathematics.* San Rafael, California: Dimensions Publishing Co.

Bever, T. H. 1970. The cognitive basis for linguistic structures. In *Cognition and the development of language,* ed. J. R. Hayes, pp. 279–352. New York: Wiley.

Bever, T. H., J. Mehler, and J. Epstein. 1968. What children do in spite of what they know. *Science* 162:921–925.

Bower, T. G. R. 1971. The object in the world of the infant. *Scientific American* 225:30–38.

Braine, M. D. S. 1962. Piaget on reasoning: a methodological critique and alternative proposal. In *Thought in the young child,* ed. W. Kessen and C. Kuhlman. *Monographs for the Society for Research in Child Development* 27, no. 2, pp. 41–63.

Brainerd, C. J. 1973. Mathematical and behavioral foundations of number. *Journal of General Psychology* 88:221–281.

Brainerd, C. J., and T. W. Allen. 1971. Experimental inductions of the conservation of "first order" quantitative invariants. *Psychological Bulletin* 75:128–144.

Brindley, G. S., and P. A. Merton. 1960. The absence of position sense in the human eye. *Journal of Physiology* 153:127–130.

Brown, A. L. 1975a. The development of memory: knowing, knowing about knowing and knowing how to know. In *Advances in child development and behavior,* vol. 10, ed. H. W. Reese, pp. 104–152. New York: Academic Press.

Brown, A. L. 1975b. Recognition, reconstruction, and recall of narrative sequences by preoperational children. *Child Development* 46:156–166.

Brown, R. 1973. *A first language: the early stages.* Cambridge, Massachusetts: Harvard University Press.

Brownell, W. A. 1941. *Arithmetic in grades I and II: a critical summary of new and previously reported research.* Durham, North Carolina: Duke University Press.

Bruner, J. S. 1957. On perceptual readiness. *Psychological Review* 64:123–152.

Bruner, J. S., et al. 1966. *Studies in cognitive growth.* New York: Wiley.

Bryant, P. E. 1974. *Perception and understanding in young children: an experimental approach.* New York: Basic Books.

Bryant, P. E., and T. R. Trabasso. 1971. Transitive inferences and memory in young children. *Nature* 232:456–458.

Bullock, M., and R. Gelman. 1977. Numerical reasoning in young children: the ordering principle. *Child Development* 48:427–434.

Carey, S., and R. Diamond. 1977. From piecemeal to configurational representation of faces. *Science* 195:312–314.

Chi, M. T. H., and D. Klahr. 1975. Span and rate of apprehension in children and adults. *Journal of Experimental Child Psychology* 19:434–439.

Chomsky, C. 1969. *The acquisition of syntax in children from 5 to 10.* Cambridge, Massachusetts: MIT Press.

Chomsky, N. 1965. *Aspects of the theory of syntax.* Cambridge, Massachusetts: MIT Press.

Churchill, E. M. 1961. *Counting and measuring.* Toronto: University of Toronto Press.

Clark, E. 1973. What's in a word? On the child's acquisition of semantics in his first language. In *Cognitive development and the acquisition of language,* ed. T. E. Moore, pp. 65–110. New York: Academic Press.

Cole, M., and S. Scribner. 1974. *Culture and thought: a psychological introduction.* New York: Wiley.

Crane, N. L., and L. E. Ross. 1967. A developmental study of attention to cue redundancy following discrimination learning. *Journal of Experimental Child Psychology* 5:1–15.

Cutting, J. E. 1977. The magical number two and the natural categories of speech and music. In *Tutorial essays in psychology,* ed. N. S. Sutherland, pp. 1–33. Hillsdale, New Jersey: Erlbaum.

Cutting, J. E., and P. D. Eimas. 1975. Phonetic feature analyzers and the processing of speech in infants. In *The role of speech in language,* ed. J. F. Kavanagh and J. E. Cutting, pp. 127–148. Cambridge, Massachusetts: MIT Press.

Dantzig, T. 1967. *Number: the language of science.* New York: The Free Press.

Descoeudres, A. 1921. *Le développement de l'enfant de deux à sept ans.* Paris: Delachaux et Niestlé C. A.

Donaldson, M., and G. Balfour. 1968. Less is more: a study of language comprehension in children. *British Journal of Psychology* 59:461–471.

Donaldson, M., and R. J. Wales. 1970. On the acquisition of some relational terms. In *Cognition and the development of language,* ed. J. R. Hayes, pp. 235–268. New York: Wiley.

Eimas, P. D. 1974. Linguistic processing of speech by young infants. In *Language perspectives: acquisition, retardation, and intervention,* ed. R. L. Schiefelbusch and L. L. Lloyd, pp. 55–73. Baltimore, Maryland: University Park Press.

Elkind, D. 1967. Piaget's conservation problems. *Child Development* 38:15–27.

Elkind, D., R. R. Koegler, and E. Go. 1964. Studies in perceptual development. *Child Development* 35:81–90.

Flavell, J. H. 1963. *The developmental psychology of Jean Piaget.* Princeton, New Jersey: Van Nostrand.

Flavell, J. H. 1970. Developmental studies of mediated memory. In *Advances in child development and behavior,* ed. H. W. Reese and L. P. Lipsitt, vol. 5, pp. 192–211. New York: Academic Press.

Flavell, J. H. 1971. Stage-related properties of cognitive development. *Cognitive Psychology* 2:421–453.

Flavell, J. H. 1977. *Cognitive development.* Englewood Cliffs, New Jersey: Prentice-Hall.

Flavell, J. H., D. H. Beach, and J. M. Chinsky. 1966. Spontaneous verbal rehearsal in a memory task as a function of age. *Child Development* 37:283–299.

Flavell, J. H., A. G. Friedrichs, and J. D. Hoyt. 1970. Developmental changes in memorization processes. *Cognitive Psychology* 1:324–340.

Flavell, J. H., and J. Wohlwill. 1969. Formal and functional aspects of cognitive development. In *Studies in cognitive development: essays in honor of Jean Piaget,* ed. D. Elkind and J. H. Flavell, pp. 67–120. New York: Oxford University Press.

Fodor, J. A. 1972. Some reflections on L. S. Vygotsky's *Thought and language. Cognition* 1:83–95.

Gallistel, C. R. Forthcoming. *The organization of action.* Hillsdale, New Jersey: Erlbaum.

Gardner, M. 1977. The concept of negative numbers and the difficulty of grasping it. *Scientific American* 236 (April): 131–134.

Gast, H. 1957. Der Umgang mit Zahlen und Zahlgebilden in der frühen Kindheit. *Zeitschrift für Psychologie* 161:1–90.

Gelman, R. 1969. Conservation acquisition: a problem of learning to attend to relevant attributes. *Journal of Experimental Child Psychology* 7:167–186.

Gelman, R. 1972a. The nature and development of early number concepts. In *Advances in child development and behavior,* vol. 7, ed. H. W. Reese, pp. 115–167. New York: Academic Press.

Gelman, R. 1972b. Logical capacity of very young children: number invariance rules. *Child Development* 43:75–90.

Gelman, R. 1977. How young children reason about small numbers. In *Cognitive Theory,* vol. 2, ed. N. J. Castellan, D. B. Pisoni, and G. R. Potts, pp. 219–238. Hillsdale, New Jersey: Erlbaum.

Gelman, R. 1978. Cognitive development. *Annual Review of Psychology* 29:297–332.

Gelman, R., and M. Shatz. 1977. Appropriate speech adjustments: the operation of conversational constraints on talk to two-year-olds. In *Interaction, conversation, and the development of language,* ed. M. Lewis and L. A. Rosenblum, pp. 27–61. New York: Wiley.

Gelman, R., and M. F. Tucker. 1975. Further investigations of the young child's conception of number. *Child Development* 46:167–175.

Gibson, E. J. 1969. *Principles of perceptual learning and development.* New York: Appleton-Century-Crofts.

Ginsburg, H. 1975. Young children's informal knowledge of mathematics. Unpublished manuscript, Cornell University.

Ginsburg, H. 1977. *Children's arithmetic.* New York: Van Nostrand.

Gleitman, L. R., and P. Rozin. 1973a. Teaching reading by use of a syllabary. *Reading Research Quarterly* 8:447–483.

Gleitman, L. R., and P. Rozin. 1973b. Phoenicians go home? (A response to Goodman). *Reading Research Quarterly* 8:494–501.

Gleitman, L. R., and P. Rozin. 1977. The structure and acquisition of reading I: relations between orthographies and the structure of language. In *Toward a psychology of reading: the proceedings of the CUNY conference,* ed. A. S. Reber and D. S. Scarborough, pp. 1–53. Hillsdale, New Jersey: Erlbaum.

Glucksberg, S., R. M. Krauss, and R. Weisberg. 1966. Referential communication in nursery school children: method and some preliminary findings. *Journal of Experimental Child Psychology* 3:333–342.

Grice, P. 1975. Logic and conversation. In *Syntax and semantics,* vol. 3, *Speech acts,* ed. P. Cole and J. L. Morgan, pp. 41–82. New York: Academic Press.

Groen, G. J. 1967. *An investigation of some counting algorithms for simple addition problems.* Technical Report no. 118. Stanford, California: Stanford University, Institute for Mathematical Studies in the Social Sciences.

Hagen, J. W., and G. H. Hale. 1973. The development of attention in children. In *Minnesota symposium on child development,* vol. 7, ed. A. Pick, pp. 117–140. Minneapolis: University of Minnesota Press.

Halford, G. S. 1970. A theory of the acquisition of conservation. *Psychological Review* 77:302–316.

Harlow, H. F. 1959. Learning set and error factor theory. In *Psychology: a study of science,* study 1, vol. 2, *General systematic formulations, learning and special processes,* ed. S. Koch, pp. 492–537. New York: McGraw-Hill.

Hochberg, J. E. 1964. *Perception.* Englewood Cliffs, New Jersey: Prentice-Hall.

Hunt, T. D. 1975. Early number "conservation" and experimenter expectancy. *Child Development* 46:984–987.

Ilg, F., and L. B. Ames. 1951. Developmental trends in arithmetic. *Journal of Genetic Psychology* 79:3–28.

Inhelder, B., and J. Piaget. 1964. *The early growth of logic in the child.* New York: Harper and Row.

Inhelder, B., and H. Sinclair. 1969. Learning cognitive structures. In *Trends and issues in developmental psychology,* ed. P. H. Mussen, J. Langer, and M. Covington, pp. 2–21. New York: Holt, Rinehart and Winston.

Inhelder, B., H. Sinclair, and M. Bovet. 1974. *Learning and the development of cognition.* Cambridge, Massachusetts: Harvard University Press.

Jensen, A. R. 1969. How much can we boost IQ and scholastic achievement? *Harvard Educational Review* 39:1–117.

Jensen, E. M., E. P. Reese, and T. W. Reese. 1950. The subitizing and counting of visually presented fields of dots. *The Journal of Psychology* 30:363–392.

Kaufman, E. L., M. W. Lord, T. W. Reese, and J. Volkmann. 1949. The discrimination of visual number. *American Journal of Psychology* 62:498–525.

Keeney, T. J., S. R. Cannizzo, and J. H. Flavell. 1967. Spontaneous and induced verbal rehearsal in a recall task. *Child Development* 38:953–966.

Keil, F. 1977. The role of ontological categories in a theory of semantic and conceptual development. Ph.D. dissertation, University of Pennsylvania.

Kemler, D. G. 1972. A developmental study of hypothesis-testing in discriminative learning tasks. Ph.D. dissertation, Brown University.

Kemler, D. G. 1975. Children's problem solving procedures in intentional discrimination tasks. Unpublished manuscript, University of Pennsylvania.

Kendler, H. H., and T. S. Kendler. 1962. Vertical and horizontal processes in problem solving. *Psychological Review* 69:1–16.

Kenyatta, J. 1953. *Facing Mount Kenya.* London: Secker and Warburg.

Klahr, D. 1973. Quantification processes. In *Visual information processing,* ed. W. G. Chase, pp. 3–34. New York: Academic Press.

Klahr, D., and J. G. Wallace. 1973. The role of quantification operators in the development of conservation. *Cognitive Psychology* 4:301–327.

Klahr, D., and J. G. Wallace. 1976. *Cognitive development, an information processing view.* Hillsdale, New Jersey: Erlbaum.

Kline, M. 1972. *Mathematical thought from ancient to modern times.* New York: Oxford University Press.

Knopp, K. 1952. *Elements of the theory of functions.* New York: Dover.

Kohlberg, L. 1968. Early education: a cognitive-developmental view. *Child Development* 39:1013–1062.

Kohlberg, L. 1969. Stage and sequence: the cognitive-developmental approach to socialization. In *Handbook of socialization theory and research,* ed. D. A. Goslin, pp. 347–480. Chicago: Rand McNally.

Kohnstamm, G. A. 1967. *Piaget's analysis of class-inclusion: right or wrong?* The Hague: Mouton.

Krauss, R. H., and S. Glucksberg. 1969. The development of communication. *Child Development* 40:255–266.

La Berge, D., and S. J. Samuels. 1974. Toward a theory of automatic information processing in reading. *Cognitive Psychology* 6:293–324.

Lawson, G., J. Baron, and L. Siegel. 1974. The role of number and length cues in children's quantitative judgments. *Child Development* 45:731–736.

Lempers, J. D., E. R. Flavell, and J. H. Flavell. 1977. The development in very young children of tacit knowledge concerning visual perception. *Genetic Psychology Monographs* 95:3–53.

Liberman, I. Y., D. Shankweiler, A. M. Liberman, C. Fowler, and F. W. Fischer. 1977. Phonetic segmentation and recoding in the beginning reader. In *Toward a psychology of reading: the proceedings of the CUNY conference,* ed. A. S. Reber and D. L. Scarborough, pp. 207–225. Hillsdale, New Jersey: Erlbaum.

Luria, A. R. 1961. *The role of speech in the regulation of normal and abnormal behavior.* New York: Pergamon.

Maccoby, E. E. 1969. The development of stimulus selection. In *Minnesota symposia on child psychology,* vol. 3, ed. J. P. Hill, pp. 68–96. Minneapolis: University of Minnesota Press.

Macnamara, J. 1974. A note on Piaget and number. *Child Development* 46:424–429.

Mandler, G., and Z. Pearlstone. 1966. Free and constrained concept learning and subsequent recall. *Journal of Verbal Learning and Verbal Behavior* 5:126–131.

Markman, E. M. 1973. Factors affecting the young child's ability to monitor his memory. Ph.D. dissertation, University of Pennsylvania.

Markman, E. M., and J. Siebert. 1976. Classes and collections: internal organization and resulting holistic properties. *Cognitive Psychology* 8:561–577.

Marler, P. 1977. Sensory templates, vocal perception, and development. In *Interaction, conversation, and the development of language,* ed. M. Lewis and L. A. Rosenblum, pp. 95–114. New York: Wiley.

Mehler, J., and T. G. Bever. 1967. Cognitive capacity of very young children. *Science* 158:141–142.

Menninger, K. 1969. *Number words and number symbols.* Cambridge, Massachusetts: MIT Press.

Merkin, S., and R. Gelman. 1978. Strategic behavior in preschoolers who are tested with a modified counting task. Unpublished manuscript, University of Pennsylvania.

Miller, P. H., and R. F. West. 1976. Perceptual supports for one-to-one correspondence in the conservation of number. *Journal of Experimental Child Psychology* 21:417–424.

Moely, B. E., F. A. Olson, T. G. Halwes, and J. H. Flavell. 1969. Production deficiency in young children's clustered recall. *Developmental Psychology* 1:26–36.

Neisser, U. 1966. *Cognitive psychology.* New York: Appleton-Century-Crofts.

Osherson, D. N., and E. M. Markman. 1974–1975. Language and the ability to evaluate contradictions and tautologies. *Cognition* 3:213–226.

Palermo, D. S. 1973. More about less: a study of language comprehension. *Journal of Verbal Learning and Verbal Behavior* 12:211–221.

Piaget, J. 1926. *The language and thought of the child.* New York: Meridian.

Piaget, J. 1928. *Judgment and reasoning in the child.* New York: Harcourt, Brace.

Piaget, J. 1929. *The child's conception of the world.* London: Routledge and Kegan Paul.

Piaget, J. 1952. *The child's conception of number.* New York: Norton.

Piaget, J. 1968. Quantification, conservation and nativism. *Science* 162:976–979.

Piaget, J. 1971. *Biology and knowledge.* Chicago: University of Chicago Press.

Piaget, J. 1975. On correspondences and morphisms. Paper presented at the Jean Piaget Society, Philadelphia, Pennsylvania.

Potter, M. C., and E. I. Levy. 1968. Spatial enumeration without counting. *Child Development* 39:265–273.

Pufall, P. B., R. E. Shaw, and A. Syrdal-Lasky. 1973. Development of number conservation: an examination of some predictions from Piaget's stage analysis and equilibration model. *Child Development* 44:21–27.

Raphael, B. 1976. *The thinking computer: Mind inside matter.* San Francisco: W. H. Freeman.

Reese, H. W. 1962. Verbal mediation as a function of age levels. *Psychological Bulletin* 59:502–509.

Reiss, A. 1943. An analysis of children's number responses. *Harvard Educational Review* 13:149–162.

Riley, C. A. 1976. The representation of comparative relations and the transitive inference task. *Journal of Experimental Child Psychology* 22:1–22.

Riley, C. A., and T. Trabasso. 1974. Comparatives, logical structures and encoding in a transitive inference task. *Journal of Experimental Child Psychology* 17:187–203.

Rosch, E., and C. B. Mervis. 1975. Family resemblances: studies in the internal structure of categories. *Cognitive Psychology* 7:573–605.

Rosch, E., C. B. Mervis, W. D. Gray, D. M. Johnson, and P. Boyes-Braem. 1976. Basic objects in natural categories. *Cognitive Psychology* 8:382–439.

Rothenberg, B. B. 1969. Conservation of number among four- and five-year-old children: some methodological considerations. *Child Development* 40 383–406.

Rozin, P. 1976. The evolution of intelligence and access to the cognitive unconscious. In *Progress in psychobiology and physiological psychology,* vol. 6, ed. J. M. Sprague and A. A. Epstein, pp. 245–281. New York: Academic Press.

Rozin, P., and L. R. Gleitman. 1977. The structure and acquisition of reading II: The reading process and the acquisition of the alphabetic principle. In *Toward a psychology of reading: the proceedings of the CUNY conference,* ed. A. S. Reber and D. S. Scarborough, pp. 55–141. Hillsdale, New Jersey: Erlbaum.

Sachs, J., and J. Devin. 1976. Young children's styles in social interaction and role playing. *Journal of Child Language* 3:81–98.

Saltzman, I. J., and W. R. Garner. 1948. Reaction time as a measure of span of attention. *Journal of Psychology* 25:227–241.

Saxe, G. B. 1977. Are counting skills necessary for the development of number conservation concepts? Unpublished manuscript, Children's Hospital Medical Center, Boston, Massachusetts.

Schaeffer, B., V. H. Eggleston, and J. L. Scott. 1974. Number development in young children. *Cognitive Psychology* 6:357–379.

Shatz, M. 1973. *Preschoolers' ability to take account of others in a toy selection task.* Masters thesis, University of Pennsylvania.

Shatz, M., and R. Gelman. 1973. The development of communication skills. *Monograph of the Society for Research in Child Development* 38, serial no. 152.

Siegel, L. 1973. The role of spatial arrangement and heterogeneity in the development of numerical equivalence. *Canadian Journal of Psychology* 27:351–355.

Siegel, L. 1974. Development of number concepts: ordering and correspondence operations and the role of length cues. *Developmental Psychology* 10:907–912.

Siegel, L. 1976. A lot about *big, little,* and *some:* the relationship between quantity discrimination and the comprehension and production of language. Unpublished manuscript, McMaster University.

Smith, F. 1971. *Understanding reading.* New York: Holt, Rinehart and Winston.

Smither, S. J., S. S. Smiley, and R. Rees. 1974. The use of perceptual cues for number judgment by young children. *Child Development* 45:693–699.

Trabasso, T. R. 1975. Representation, memory and reasoning: how do we make transitive inferences? In *Minnesota symposium on child psychology,* vol. 9, ed. A. D. Pick, pp. 135–172. Minneapolis: University of Minnesota Press.

Trabasso, T. R., and G. Bower. 1968. *Attention in learning: theory and research.* New York: Wiley.

Trabasso, T. R., J. A. Deutsch, and R. Gelman. 1966. Attention in discrimination learning of young children. *Journal of Experimental Child Psychology* 4:9–19.

Turiel, E. 1969. Development processes in the child's moral thinking. In *Trends and issues in developmental psychology,* ed. P. H. Mussen, J. Langer, and M. Covington, pp. 92–133. New York: Holt, Rinehart and Winston.

Vygotsky, L. S. 1962. *Thought and language.* Cambridge, Massachusetts: MIT Press.

Wallach, L., and R. L. Sprott. 1964. Inducing number conservation in children. *Child Development* 35:1057–1071.

Wallach, L., A. J. Wall, and L. Anderson. 1967. Number conservation: the roles of reversibility, addition-subtraction, and misleading perceptual cues. *Child Development* 38:425–442.

Wang, M. C., L. B. Resnick, and R. F. Boozer. 1971. The sequence of development of some early mathematics behaviors. *Child Development* 42:1767–1778.

Weinstein, M. L. 1977. Subitizing? Probably not: evidence for an automatized counting model for perception of the numerosity of small arrays. Unpublished manuscript, University of Pennsylvania.

Weir, R. H. 1962. *Language in the crib.* The Hague: Mouton.

Werner, H. 1957. *The comparative psychology of mental development,* 2nd ed. New York: International Universities Press.

White, S. H. 1965. Evidence for a hierarchical arrangement of learning processes. In *Advances in child development and behavior,* vol. 2, ed. L. P. Lipsitt and C. C. Spiker, pp. 187–220. New York: Academic Press.

Wilkinson, A. 1976. Counting strategies and semantic analysis as applied to class inclusion. *Cognitive Psychology* 8:64–85.

Woodworth, R. S., and H. Schlosberg. 1954. *Experimental psychology.* New York: Holt.

Worden, P. E. 1974. The development of the category-recall function under three retrieval conditions. *Child Development* 45:1054–1059.

Worden, P. E. 1975. Effects of sorting on subsequent recall of unrelated items: a developmental study. *Child Development* 46:687–695.

Worden, P. E. 1976. The effects of classification structure on organized free recall in children. *Journal of Experimental Child Psychology* 22:519–529.

Wundt, W. 1907. *Outlines of psychology.* Leipzig: Wilhelm Engelmann. New York: G. E. Stechert and Co.

Zaslavsky, C. 1973. *Africa counts.* Boston: Prindle, Weber and Schmidt.

Index

1491